The Functional Consequences
of Biodiversity

MONOGRAPHS IN POPULATION BIOLOGY

EDITED BY SIMON A. LEVIN AND HENRY S. HORN

Titles available in the series (by monograph number)

The Functional Consequences of Biodiversity

Empirical Progress and Theoretical Extensions

ANN P. KINZIG,
STEPHEN W. PACALA,
AND
DAVID TILMAN,
EDITORS

PRINCETON UNIVERSITY PRESS

PRINCETON AND OXFORD

2001

Library of Congress Control Number 2001097325

ISBN: 0-691-08821-7 (hardcover); 0-691-08822-5 (paper)
British Library Cataloging-in-Publication Data is available

This book has been composed in Baskerville

Printed on acid-free paper. ∞

www.pup.princeton.edu

Printed in the United States of America

10 9 8 7 6 5 4 3 2 1

10 9 8 7 6 5 4 3 2 1
(Pbk.)

Dedicated to
Ernst-Detlef Schulze and Harold A. Mooney

Contents

CONTENTS

x

CONTENTS

xi

Preface

Ideas that lay dormant for 20 or more years came to life with the 1993 publication of *Biodiversity and Ecosystem Function*. Edited by Detlef Schulze and Hal Mooney, and inspired by a conference they led in Bayreuth, Germany, this book used the theoretical tools, experimental results, and observational data that had accumulated in the intervening decades to re-explore questions originally raised by some of the great minds of the discipline, including Charles Darwin (1858), Charles Elton (1958), Eugene Odum (1953), and Robert May (1974).

The eight years since its publication have been a period of unprecedented activity and controversy. Coming in rapid succession were experiments in controlled-environment settings (e.g., Naeem et al. 1994, 1995; McGrady-Steed, Harris, and Morin 1997), analyses of experiments started for other reasons (e.g., Tilman and Downing 1994; Tilman 1996), large-scale biodiversity field experiments (e.g., Tilman, Wedin, and Knops 1996, 1997a)—including a biodiversity experiment replicated across eight European nations (Hector et al. 2000)—and observational studies (e.g., Wardle et al. 1997a). These papers, and a large number of subsequent treatments that are reviewed in this book, contributed major new insights into the effects of biodiversity on ecosystem productivity, stability, and nutrient dynamics. They also ignited major controversies. These insights and controversies are the bases for the syntheses presented in this book.

Controversy began with the first new papers on biodiversity and ecosystem functioning. Givnish (1994) objected to the interpretation of the Cedar Creek drought study (Tilman and Downing 1994), asserting that factors other than biodiversity might be responsible for the apparent effect of biodiversity on ecosystem stability. Huston (1997) sided with Givnish (1994). Doak et al. (1998) proposed an alternative way for greater biodiversity to give greater ecosystem stability. In addition, Huston (1997) asserted that the Ecotron foodweb experiment (Naeem et al. 1994) had fatal flaws in its experimental design, and that the results of the Cedar Creek biodiversity field experiments could have come not from biodiversity but from a 'hidden treatment'—the greater chance that productive species would be present at higher diversity. Aarssen (1997) independently proposed this 'sampling effect' as an explanation for the apparent effects of diversity on productivity. In response to a comparative study of ecosystem processes on islands by Wardle (1997a) and a report on a second Cedar Creek biodiversity experiment by Tilman et al. (1997a), Grime (1997) expressed concerns about the interpretation of biodiversity experiments and their relevance to patterns in natural communities. These are highlights of what, for ecology, has been unusually rapid-fire, high-profile research, with many of the experimental papers, the criticisms of them, and the responses to these criticisms, appearing in *Science* and *Nature*.

Why has there been such controversy? The assertion that biodiversity could be one of the factors controlling ecosystem functioning represented a paradigm shift in ecology. Biodiversity, which had long been thought to be controlled by productivity and environmental variation, was now being suggested as controlling productivity and ecosystem variability. This reversal of long-held cause-and-effect relationships forced re-exploration of concepts and definitions, and demanded the development of new hypotheses. Indeed, much of the controversy was debate about what might be causing biodiversity to influence various ecosystem processes. The initial ideas—idiosyncratic responses, the rivet hypothesis, curves of one of three shapes, etc.—were much more phenomenological and descriptive than mechanistic. These initial ideas have been replaced, in an intellectual successional sequence typical of a newly developing discipline, by a diversity of increasingly mechanistic explanations. The ideas of Givnish (1994), Huston (1997), Grime (1997), Aarssen (1997), Tilman, Lehman, and Thomson (1997), Doak et al. (1998), and others have provided a rich array of possibilities against which the results of biodiversity experiments and observational studies can be tested. These new tests, in turn, suggest other ways that diversity may impact ecosystem processes. This book presents such tests and suggests new explanations and new theory.

A second factor fueled interest in this topic, and contributed to the controversy. Because of human-caused global change, ecological research now often addresses issues that are not only of academic interest, but also of direct relevance to public policy. The initial studies demonstrating links between biodiversity and ecosystem productivity and stability were thus widely covered by the media, as were subsequent results. This necessarily broadened the debate to include discussion of not just the science but also of the relevance of the science to public policy.

All of this was swirling around when we met at the National Center for Ecological Analysis and Synthesis (NCEAS) at UC Santa Barbara at the end of 1999. Our goals were to present our most recent experimental and theoretical results and to hold a free-ranging dialog on them. We sought to summarize and synthesize what we knew empirically, and to find a deeper conceptual and theoretical understanding of what all this implied for the relationships between biodiversity and ecosystem functioning. Based on these discussions, we revised the thoughts expressed in the initial drafts of our chapters, and also commissioned several new chapters to develop ideas that had emerged at NCEAS.

All of those attending the NCEAS meeting were fully aware of the controversies that had accompanied biodiversity work. However, nothing had led any of us to anticipate that, within six months, the biodiversity and ecosystem functioning controversy would explode. The publication of a letter by Wardle et al. (2000) in the *Bulletin* of the Ecological Society of America, and subsequent coverage of this by *Science* (Kaiser 2000) and the *Chronicle of Higher Education* (Guterman, 2000), raised the profile of this debate to new heights. Who would have thought that *Science* would have a headline titled "Rift Over Biodiversity Divides Ecologists," and a subhead of "An acrimonious dispute has broken out over whether the data on biodiversity are robust enough to inform public policy" (Kaiser 2000)?

We make no claims that this book resolves this controversy. Indeed, it was not written to explicitly address this controversy. However, all of the chapters in this book were influenced by the debate about biodiversity and ecosystem functioning. This debate was an ever-present challenge, forcing deeper exploration, and encouraging synthesis. Although additional experiments, additional theory, and greater consensus on terminology are needed, the results, summaries, and syntheses offered in this book are, we hope, the next logical step in our understanding of the effects of diversity on ecosystem functioning.

This book was inspired by Diversitas and by the Global Change in Terrestrial Ecosystems program (GCTE), and represents the confluence of two initially disparate efforts. Ann Kinzig and Steve Pacala began organizing, as part of Focus 4 of the GCTE initiative, a meeting on theory relevant to biodiversity and ecosystem functioning. Osvaldo Sala and David Tilman simultaneously began a GCTE-Focus 4 effort to synthesize experimental results and discuss designs for new experimental treatments of diversity and ecosystem functioning relationships. Two proposals went to NCEAS, but one meeting resulted—a meeting that pursued both goals simultaneously. We are deeply indebted to NCEAS and Jim Reichman. Osvaldo Sala contributed greatly to the meeting and to this book. We thank Sam Elworthy of Princeton University Press for support and encouragement, and the staff of NCEAS for their logistic support.

We reserve our greatest thanks to all who have contributed their efforts and ideas—especially to those younger ecologists who have pursued this issue, despite the controversy, precisely because of its importance both to the discipline and to public policy. Although acknowledged via the usual method of scientific citation throughout the chapters of this book, added recognition is due both to those who have done experimental and observational studies, knowing that their papers would be subject to deep scrutiny, and to those who have offered novel, at times controversial, ideas and interpretations. The willingness to do each of these is essential to healthy science.

There is much more to be done on biodiversity and ecosystem functioning. And, there are a large number of other issues where fundamental ecological research is also relevant to public policy. We hope that the work in this book might help inspire a new generation of ecologists to pursue the greatest challenges our discipline can offer—those that are of importance to both science and society. Those who do so can anticipate an exhilarating career.

Contributors

Juan Armesto
  Universidad de Chile

Teri Balser
  Stanford University

Peter Chesson
  University of California,
  Davis

Mary K. Firestone
  University of California,
  Berkeley

Andy Hector
  Imperial College at
  Silwood Park

Robert Holt
  University of Kansas

Jasmin Joshi
  University of Zurich

Peter Kareiva
  National Marine Fisheries
  Service
  National Oceanic and
  Atmospheric
  Administration

Ann P. Kinzig
  Arizona State University

Johannes Knops
  University of Nebraska

Sharon Lawler
  University of California,
  Davis

Clarence Lehman
  University of Minnesota

Michel Loreau
  Ecole Normale Superieure

Shahid Naeem
  University of Washington

Claudia Neuhauser
  University of Minnesota

Stephen Pacala
  Princeton University

Peter Reich
  University of Minnesota

Felix Schläpfer
  University of Zurich

Bernhard Schmid
  University of Zurich

David Tilman
  University of Minnesota

David Wedin
  University of Nebraska

Figures

Tables

CHAPTER ONE

Opening Remarks

Ann P. Kinzig

Darwin first proposed a connection between biodiversity and ecosystem functioning in 1859. Interest in the topic has mainly waned—sometimes waxed—in the time interval since. In the last few decades, however, the waxing has had the upper hand, and a quick trip through an electronic archive reveals over 100 articles on this topic since 1982.

Why then a book on the subject? First, there have been significant advances in our empirical understanding of the diversity–functioning relationship in the last few years, but those results have not been compiled, evaluated, and synthesized in both a comprehensive and detailed manner elsewhere. Second, we offer new theoretical results that advance our understanding of when and under what circumstances certain forms of the diversity–functioning relationship might emerge. Third, while knowledge of consumer and decomposer influences on diversity–functioning relationships has not advanced in the last decade as much as many of us had hoped it would, some sensible recommendations for focusing research efforts can be made, and we offer those here. Perhaps most importantly, our analysis of the existing experimental record yielded some surprising results, and our attempts to explain those results and extend them with development of new theory significantly influenced our perceptions of the mechanisms governing diversity and ecosystem functioning relationships.

This book, then, is organized into three parts—a synthesis of the existing experimental and theoretical work and the

1

interpretation of each in light of the other, new contributions to theory that extend the experimental results, and a discussion of future directions.

The first section leads with a chapter by Tilman and Lehman that presents the current prevailing theories of the mechanisms governing the diversity–functioning relationship, and the evidence for or against these mechanisms in early experimental results. The next two chapters—by Tilman et al. (chapter 3) and Hector (chapter 4)—give detailed analyses of the two most extensive diversity–functioning experiments to date, in the grasslands of Minnesota and Europe, respectively. Naeem (chapter 5) reviews the literature on trophic interactions and microbial influences on ecosystem processes, and offers suggestions for advancing the field using both theoretical and empirical approaches. In chapter 6, Schmid, Joshi, and Schläpfer offer a comprehensive review of experimental and observational studies that should serve as one of the most useful compilations of approaches and results for any serious scholar in this field. Pacala and Tilman (chapter 7) conclude this section with a discussion of the biggest surprises that have emerged in the existing experimental record, and provide some simple explanations for these surprises.

In spite of the advances made in the experimental record, however, practical limitations simply preclude experiments that can span the large spatial scales, the long temporal scales, and the representative diversity gradients and structures that are properly the concern of work in this area. Thus we use the next section, on theory, to extend diversity–functioning results to scales that are not particularly amenable to experimental manipulations—landscape-level (as opposed to plot-level) processes, the long time dynamics that govern and characterize natural systems, and multitrophic-level interactions. In doing so, we take a very specific approach. In particular, we employ a common ecosystem model and "attach" to that common model different models of species coexistence. We can thus attribute differences in outcome to differences in coexistence mechanisms, rather than to differences in approaches taken for simulating decomposition, mineralization, and so forth. The common model is general enough to accommodate many of the ecosystem processes of interest, and is presented in chapter 8. Kinzig and Pacala (chapter 9) then look at "successional niche" models—processes of succession in lightly disturbed environments or competition-colonization dynamics in more heavily disturbed environments. Chesson, Pacala, and Neuhauser (chapter 10) examine diversity–functioning relationships when coexistence is maintained by spatial or temporal heterogeneity—a patchwork of soil types, or changing climatic conditions throughout seasons, for instance. Finally, Holt and Loreau (chapter 11) examine simple models of trophic interactions and contrast their results with the previous chapters, where only the plant community was explicitly included in the ecosystem model.

In the last section of the book Balser, Kinzig, and Firestone (chapter 12) sift the literature on the influences of microbial diversity on ecosystem functioning, and offer a framework for guiding future work in this area. Lawler, Armesto, and Kareiva (chapter 13) examine the implications of these scientific findings for conservation. Finally, in the last chapter, the editors offer a synthesis of the major conclusions, and a blueprint for future research directions.

This book is not, however, the last word on the subject of diversity–functioning relationships. In fact, the characterization of any vigorous scientific field is that obsolescence begins the moment a project is conceived. So too with our efforts. In particular, when we developed the approach used in the theory section, well over three years ago now, the early experimental results all seemed consistent with a sampling mechanism—that is, dominance of polycultures by a single best competitor, and a best polyculture performance that equaled, but did not exceed, the best monoculture performance.

We now see experimental results that differ from these—a "transition," if you will, between a sampling mechanism to niche-partitioning or niche-complementarity mechanisms. These niche mechanisms permit an "overyielding" effect—where the best polyculture can outperform the best monoculture. This observed transition justifies some of the early criticisms of Huston (1997), Aarssen (1997), and others—namely, that what we were seeing in the initial experimental results was domination by the "weediest," fastest-growing species, a dynamic that has little to do with any interesting features of diversity–functioning relationships. These experimental outcomes also force us to revisit the earlier analyses of MacArthur, where just such niche complementarity and overyielding were predicted to occur.

Nonetheless, the early experimental results, and the overly harsh discreditation of MacArthur's work, led us to focus on possible sampling mechanisms over niche-complementarity mechanisms. Thus, while we include some complementarity mechanisms in the theory section (see, particularly, chapter 10), we overlook other classes of models that might have proved useful in interpreting observed relationships and extending experimental results. We return to this point in the concluding chapter.

In retrospect, we should not have been surprised by this transition from sampling to niche complementarity. As ecologists, we know that systems away from equilibrium (as planted plots in experimental manipulations certainly are) can be characterized by two (or more) distinctive time dynamics—an initial, rapid response that has to do with early growth dynamics, and a longer-term response governed by competitive interactions and population dynamics. (Pacala and Tilman elaborate on these points, and offer formal mathematical representations, in chapter 7.) Thus, we should have expected potentially significant changes in outcomes as time wore on. Nor should we have been surprised by the emergence of niche complementarity—we know that species coexist, and we understand that this coexistence can mean changes in performance that do not "balance out" between monocultures and polycultures.

In addition, the analyses presented in this book illuminate some conspicuous exceptions to the generally expected result that ecosystem functioning will increase with diversity, and that the performance of the best polycultures will equal or exceed the performances of the best, but more depauperate, communities. In particular, there is some experimental evidence for declining functioning with increasing diversity, and this same pattern emerges from the theories developed for communities where coexistence is maintained by successional niches or interference competition. These outcomes can be understood by considering the relationship between competitive ability and the performance of certain ecosystem processes. Ultimately, the influence of diversity per se on ecosystem functioning should depend on those mechanisms permitting coexistence in particular systems, and on how those strategies that permit competitive superiority or dominance in these coexistence battles influence performance at both local and landscape scales. For instance, are competitive rank and performance of certain functions "parallel," with the superior competitor being the superior performer? Or are they "orthogonal," with no discernible relationship between competitive ability and performance? Might a superior competitor have high local performance but low landscape performance due to low abundance and suppression of other types? How do these relationships between competition and performance depend on scale, or on the coexistence mechanism operating, or on the ecosystem process being examined?

These emerging insights into the more complicated dynamics and mechanisms governing diversity–functioning relationships provide one of the most compelling reasons for further study in this area—namely, that disentangling the influences of diversity on ecosystem functioning, across space and time, requires that we deepen our understanding of the answers to some of the most compelling questions in ecology. How do species coexist? Why do we see so many together, and why don't we see more? How do insights into patterns and processes derived at one spatial and temporal scale translate across other scales? How do communities assemble, how do they disassemble, and how do these rules change when humans decide to rearrange their surroundings? (See Levin 1999 for a wonderful exposition on many of these questions.) Ultimately, then, the study of biodiversity–ecosystem functioning relationships is no more and no less than the study of some of the most fundamental questions in ecology.

Finally, though, we are citizens as well as scientists, with an interest in sustaining critical ecosystem processes and species integrity, which are under increasing assault from the sheer magnitude of the human endeavor. We therefore cannot ignore the social, ethical, and political implications of our work in this field—what it might mean for the design and control of managed systems, for the likely responses of ecosystems to global change and the policies needed to mediate those responses, for the choices we must make in the species triage we conduct every day. We only touch on some of those implications in this volume, but as scientists and as citizens we are acutely aware of them. We do hope, after all, to leave a world that sustains human well-being, and to leave a world in which tomorrow's scientists can answer the questions we fail to answer today.

PART 1

Empirical Progress

CHAPTER TWO

Biodiversity, Composition, and Ecosystem Processes: Theory and Concepts

David Tilman and Clarence Lehman

INTRODUCTION

Although Darwin (1859) hypothesized that diversity should affect productivity, and cited agricultural evidence in support of this assertion (see McNaughton 1993), further interest in the potential effects of diversity on stability, productivity, and invasibility began with Odum (1953), MacArthur (1955), Margalef (1969), and especially Elton (1958). These individuals, and others of that era, offered a variety of reasons why the rates of various community and ecosystem processes, especially those related to stability and community invasibility, might depend on diversity. As was the tradition of the era, these concepts were mainly developed using verbal "models," and the field evidence cited in support of the concepts came from informal comparisons of habitats (e.g., of tropical versus temperate habitats, of islands versus mainland, or of farmland versus native vegetation). In contrast, Robert May (1972) introduced rigorous mathematical treatments of a variety of population ecology questions, including the relationship between diversity and stability. As he points out in the rerelease of his 1973 book by Princeton University Press as an "Ecological Classic" (May 2001), the effects of diversity were not his central focus, and, indeed, the word *diversity* did not even appear in the index. Nevertheless, his models of multispecies competition showed that species were less stable when living in communities with a greater diversity of competitors, and that the effects of diversity on population stability were small and subtle. The rigor of May's (1972) mathematics, the paucity of rigorous experimental or observational evidence, and the loose super-organismal language in which some of the earlier ideas had been framed led Goodman (1975), in a major review, to conclude that there likely was no clear relationship between diversity and stability. McNaughton (1978), Pimm (1979, 1984), Ehrlich and Ehrlich (1981), King and Pimm (1983), and others objected to Goodman's conclusions and continued to explore the issue. Nonetheless, the issue faded from the forefront of ecological research, with many ecologists considering diversity to be of little importance to stability and other community and ecosystem processes.

Ehrlich and Ehrlich (1981), Wilson (1992), and especially the papers in Schulze and Mooney (1993) renewed interest in the potential effects of biodiversity on various ecosystem processes. This renewed interest has raised a series of questions, including the fundamental issues of whether or not various ecosystem processes do, indeed, depend on biodiversity, and if so, how. These issues have been addressed in a recent series of laboratory and field experiments, and in field observations, which are reviewed in the chapters that follow. Equally important, though, are the conceptual bases for the potential effects of biodiversity on ecosystem functioning. Many novel concepts and mathematical theories have been proposed since Schulze and Mooney (1993). The goal of this chapter is to provide a synthesis of concepts linking biodiversity to ecosystem processes, of underlying reasons why such effects might occur, and of situations to which alternative theories may apply. These concepts can then be evaluated relative to the experimental and observational results presented in the chapters that follow, and compared to some new theory, which is presented in Pacala and Tilman (chapter 7), in Kinzig and Pacala (chapter 9), in Chesson, Pacala, and Neuhauser (chapter 10), and in Holt and Loreau (chapter 11).

DEFINITIONS OF DIVERSITY

Before discussing concepts and theories, it is important to clarify terminology. As classically defined, the species diversity of a community is based on just the number of species in that locality (i.e., species richness or species number), or on both species number and the relative abundances of these species (i.e., the Shannon index, H', its derivative $e^{H'}$, or other diversity indices), but does not take into account species composition. By themselves such definitions are inadequate for determining the potential effects of diversity on ecosystem processes because they can allow differences in species diversity to be confounded with differences in species composition. For instance, Elton (1958) discussed the apparently greater stability of diverse native vegetation than of nearby agricultural monocultures, and suggested that this difference might be caused by diversity. However, such differences in stability, should they occur, might be caused by differences in composition, or in other factors that covary among such sites. After all, it is now well known that species composition has a major impact on community and ecosystem processes (e.g., Paine 1966; Pastor et al. 1984; Vitousek et al. 1987; Wedin and Tilman 1990; Estes and Duggins 1995; Ewel and Bigelow 1996).

In order to distinguish among effects of diversity and other variables, it is necessary to refine the classical definition of diversity. In doing so, it is important to recognize that diversity has several correlated components. One classically recognized component is measured by the number of species present in a habitat, which is often called *species number*, *species richness*, or *species diversity*. This is the component

11

that has been the focus of most of the recent research on biodiversity and ecosystem functioning. A second component, called *functional diversity*, is ideally measured by the range of species traits in a habitat or in a regional or experimental species pool.

We define diversity effects as being those attributable to the number of species, or to an index based on species number and relative abundances, or to the range in the traits of the species, once there has been simultaneous control for composition and other potentially confounding variables. To distinguish effects of species diversity from those of species composition, it is necessary either (1) to experimentally hold composition "constant" via randomization (numerous communities with compositions randomly chosen from the same species pool) while changing diversity, or (2) to experimentally hold diversity constant while changing composition, or (3) to simultaneously vary both in an appropriately randomized and replicated experimental design, or (4), for observational studies, to use multiple regression or other statistical techniques to separate effects attributable to composition from those attributable to diversity. Variables other than composition that might also be confounded with diversity need to be treated similarly, especially in analyses of data gathered in observational and comparative studies.

There is an additional problem with the classical definition of diversity. Consider, for instance, two forest stands, one composed of five different, equally abundant, tree species all within the same genus (such as five different species of *Quercus*), and the other composed of five different, equally abundant, tree species, each within a different genus (e.g., *Quercus, Acer, Pinus, Fagus,* and *Liriodendron*). Which stand is more diverse? A simple counting of species numbers would say that the stands had identical species numbers. The stands would also have identical species diversities if measured using H' or $e^{H'}$ or indices of evenness or equitability. And yet, almost anyone entering one stand and then the other would consider the stand containing five different genera of trees to be more diverse than the stand containing only one genus. This is so because the intuitive definition of diversity is that it measures the differences among organisms within a community. Some of these differences are captured by knowing how many species occur in the community (i.e., species number), and their relative abundances (i.e., H' and $e^{H'}$). However, other aspects of the differences occurring in communities would be captured only by a diversity index that also incorporated either the functional traits of the species or some measure of higher taxonomic diversity, such as via the number of different genera, or families, or classes, and so on, that occurred in the community, and their relative abundances. We know of no index that incorporates measures of taxonomic diversity, and suggest that such an index, perhaps based on phylogenetic differences, might provide a useful way to measure the effects of the range of species traits.

Taxonomic diversity might provide a reasonable measure of the differences among organisms in a community. However, a variety of nontaxonomic diversity indices might be able to provide equally good measures of such differences. Vitousek and Hooper (1993), Lawton and Brown (1993), and others have suggested that an alternative way to measure diversity is by determining the range of functional traits represented in a community. One way to do this is to classify organisms into functional groups (e.g., C3 plant species, C4 plant species, grasses, legumes, annuals, perennials, etc., and various combinations of such traits) and to use the number of functional groups in a community, termed *functional group number*, as a measure of its diversity. This could be expanded to include the relative abundances of the functional groups via use of H' or $e^{H'}$, where H' would be calculated using the relative abundances of functional groups. Alternatively, the actual functional traits of each species could be determined, and the proportion of a multidimen-

13

sional trait space covered by a particular suite of species could be used as a measure of the functional diversity of the community.

In total, there are a variety of ways, including species number, functional group number, and taxonomic diversity, to estimate the range of species traits that occur in a community. All of these provide measures of species diversity. Species number or functional group number have the advantage of simplicity. More detailed indices have the disadvantage of being more difficult to estimate but the advantage of being more complete measures of diversity. All such indices are likely to be highly correlated. In general, unless stated otherwise, diversity, species diversity, species number, and species richness are used interchangeably in this book to mean simply the number of species in a habitat. This is done in explicit recognition of the correlation between this simple index and the range of species traits present in a habitat.

Problems Related to Experiments and Observations

Allison (1999) discusses the strengths and weaknesses of alternative experimental designs for testing for the effects of diversity. As Allison points out, there are trade-offs among experimental designs, and an experiment that has the power to test adequately one set of diversity-related hypotheses may not have the power to test different hypotheses.

Although a major goal of ecological research is to uncover the causes of patterns in natural and managed or impacted ecosystems, it is difficult to interpret unambiguously nonexperimental field observations that may seemingly relate diversity to ecosystem functioning. This is so because the factors that cause site-to-site differences in diversity, such as habitat productivity, disturbance history, and rates of arrival of propagules, can simultaneously impact ecosystem functioning and can cause correlated changes in species composition (Loreau 1998b). For instance, diversity, composition, and other factors were highly interdependent for the field observations of Wardle et al. (1997) on a series of islands because fire frequency, which depended on island size, influenced all of these. This meant that responses that Wardle et al. (1997) attributed, albeit rather tentatively, to diversity could have been caused by associated changes in composition and disturbance. Similar correlations among disturbance history, productivity, diversity, and composition make it difficult to apply the concepts reviewed in this chapter directly to broad-scale comparisons of ecosystems, or to use broad-scale comparisons, such as patterns along latitudinal or geographic gradients, to infer the effects of diversity. As indicated by the definition of diversity that we use, the effects of diversity are most readily observed at local scales both because the effects of diversity come from the neighborhood interactions of members of different species and because most local communities are more likely to have compositional differences determined by random forces such as chance colonization than by differences in disturbance rates, soil type, and so forth. Such issues are addressed in more detail by Schmid, Joshi, and Schläpfer (chapter 6) and by Kinzig, Pacala, and Tilman (chapter 14). Here we address the effects of diversity on community productivity and resource use, and on population and community stability.

DIVERSITY, PRODUCTIVITY, AND RESOURCE DYNAMICS

Darwin (1859), as quoted by McNaughton (1993), asserted that greater diversity implied fuller exploitation of habitats and thus greater productivity:

> The more diversified in habits and structures the descendants . . . become, the more places they will be enabled to occupy. . . . It has been experimentally proved, if a plot of ground be sown with one species of grass, and a similar

plot be sown with several distinct genera of grasses, a greater number of plants and greater weight of dry herbage can be raised in the latter than the former case.

Trenbath (1974) found, in an analysis of 572 different intercropping experiments, that intercrops had about 10% greater yield than monoculture crops. Based on such evidence, Swift and Anderson (1993) suggested that the productivity of plant communities might be an increasing, but saturating, function of crop plant diversity. This hypothesized relationship hinged on greater plant diversity being associated, on average, with a greater range in plant traits related to resource capture and use. Intercropping has been found to lead to greater yields mainly when the plants have markedly different functional traits, such as for a grass (corn) and a legume (beans). If species differ in the types, amounts, or spatial or temporal pattern of resource utilization, then greater diversity is hypothesized to lead to greater utilization of limiting resources and thus to greater productivity (Ewel, Mazzarino, and Berish 1991; Vitousek and Hooper 1993). Thus, more diverse ecosystems would be expected to have lower levels of unconsumed resources.

Sampling Effect Models

These ideas were expanded and made explicit in two qualitatively different ways—in sampling effect models and in niche differentiation models. Aarssen (1997), Huston (1997), and Tilman, Lehman, and Thomson (1997) simultaneously suggested the potential role of a "sampling effect" as a cause of diversity–productivity relationships. In essence, the sampling effect model asserts that, all else being equal, any given species is more likely to be present at higher diversity, that is, it is more likely to have been "sampled" from the pool of species that occur in the region. If the presence of a particular species, such as the dominant competitor, has a great impact on the functioning of an ecosystem, as diversity increases, the system should, on average, tend to take on the functioning imposed by this species because of the greater likelihood of it being present at higher diversity. This process is one potential explanation for the higher plant productivity that had been observed in biodiversity experiments (Naeem et al. 1994, 1995; Tilman, Wedin, and Knops 1996). In particular, the plant species in an experimental species pool might differ in both their competitive abilities and productivities, with the best competitors also being the most productive species. This assumption, combined with having communities randomly assembled from a pool of species, could cause ecosystem productivity to increase with diversity. Alternatively, species could gain competitive ability not by being better exploitative competitors but by being better at interference (antagonistic) competition. If this were so, and if species that were better at competing via this mechanism were less productive, this version of the sampling effect would cause productivity to decline as diversity increased.

Let us consider a sampling effect in which individual species differ in their productivity, species compete for a single limiting resource, and species attain greater competitive ability via more efficient resource utilization (i.e., a lower R^*, *sensu* Tilman 1982), which leads to greater productivity. If the species compositions of communities are determined by random draws from a pool of species differing in their R^* values, then the likelihood that a better competitor (a species with a lower R^* value) would be present at a site would increase as diversity increased. Because these species are more productive, ecosystem productivity would increase, on average, as the initial plant diversity increased. Simulations of these assumptions lead to the curves of figure 2.1A.

This sampling effect was made mathematically explicit in a model based on the assumptions above (Tilman, Lehman, and Thomson 1997). The model provides an analytical prediction of both the mean effect of diversity on productivity

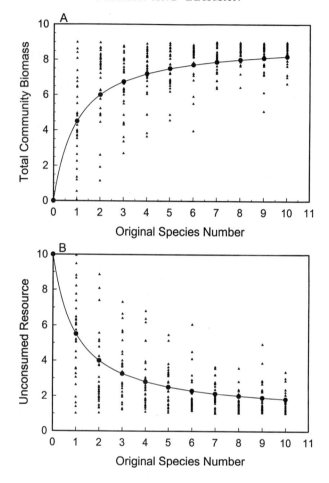

FIGURE 2.1. Sampling effect model of exploitative competition for a single limiting resource, based on Tilman, Lehman, and Thomson (1997). (A) Dependence of total community biomass at equilibrium on original species number. Triangular symbols show results of simulations of the sampling effect, large dots show means of the simulations, and the solid curve is the response predicted by equation 2.1. Note that in this version of the sampling effect, a single species is the best competitor and it displaces all other species at equilibrium. (B) Effect of initial species number on the environmental concentration of the resource, which is a direct measure of the amount of resource left unconsumed. Symbols as for part A. Figure modified from Tilman, Lehman, and Thomson (1997).

and of its variance, given the assumption that better competitors are more productive. In particular, total community biomass is predicted by the sampling model to be an increasing function of diversity, N:

$$B_{(N)} = c \left[S - \frac{2M - D}{2} - \frac{D}{N + 1} \right] \qquad (2.1)$$

where $B_{(N)}$ is the average total community biomass, at equilibrium, for a community initially containing N species ($N \geq 1$), c is a constant, S is the rate of supply of the limiting resource, M is the mean trait (mean R^* value) for the entire community, and D is the range of interspecific differences in this trait within the entire community. D is thus a direct measure of functional diversity. Here, for a uniform distribution of R^* values, $D = R^*_{max} - R^*_{min}$, where R^*_{max} is the highest R^* value and R^*_{min} is the lowest value in the uniform distribution. Note that equation 2.1 is defined for $N \geq 1$. For $N = 0$, $B = 0$.

This sampling effect model predicts that equilibrial total community biomass is an asymptotically increasing function of the initial number of species present, N (figure 2.1A), and a linearly increasing function of functional diversity, that is, of the range of species traits, D. The effect of D on total community biomass is greater if there are more species present (because $\partial B / \partial D = c(\frac{1}{2} - 1/(N + 1))$). For a given species pool (and thus for a fixed M and D), the range of species traits represented in a community depends on its N, which makes N a measure of diversity. However, if both D and N vary, diversity also depends on D. The species pool composed of five different species of *Quercus*, for instance, would likely have a smaller D than one composed of five tree species, each in a different genus. D thus measures the potential range of traits for a region, and N measures the number of these present in a given community. Both D and N contribute to the actual range of traits in a community, that is, to its diversity. Increasing either N or D, or both, is predicted to increase productivity. If $D = 0$, which means that all species have identical traits, there is no effect of N on total community biomass, a gratifying and intuitively obvious result. Similarly, when $N = 0$, $B = 0$ no matter what the value of D.

In this sampling effect model, the most productive species is also assumed to be the best competitor, and all species are assumed to be competing for a single resource. In such situations, the single species with the lowest requirement (measured as its R^*) for that resource would displace all other species at equilibrium. Thus, this sampling effect model predicts that all communities would become monocultures at equilibrium.

The response of a given randomly chosen community depends on the traits of the species it contains. This causes there to be variance around the mean response predicted in equation 2.1. This variance is totally attributable to species composition, that is, to compositional differences associated with the large number of alternative community compositions that are possible at each level of species diversity. The exact relationship for the predicted variance, given in Tilman, Lehman, and Thomson (1997), simplifies to approximately the following dependence of the standard deviation, σ, of total community biomass on species number, N:

$$\sigma_{(N)} = \frac{cD}{N + 2}.$$ (2.2)

This analysis predicts that the magnitude of the variation among communities that are started with a given number of randomly chosen species should depend directly on the interspecific range in their relevant traits (D) and inversely on initial species number (which is visually clear in figure 2.1A). Thus, greater species number is predicted to lead to lower variability among communities. In the limit, as diversity approaches infinity, the variation approaches zero,

20

which is intuitively satisfying because the species pool is in-
finitely large in this case. In two laboratory studies of the
reliability or predictability of multispecies communities,
both Naeem and Li (1997) and McGrady-Steed, Harris, and
Morin (1997) found that greater diversity did lead to lower
variation and thus to greater reliability.

The sampling effect model also predicts that the average
level of unconsumed resource, and the variance in this, decline as species number increases (figure 2.1B). Indeed,
within this formulation of the model, the greater average
total community biomass of communities started with greater
diversity is directly related to their greater consumption and
use of the limiting resource.

Inspection of figure 2.1A reveals a pattern of the dependence of the variability in total community biomass on species number that is diagnostic of the sampling effect: the
upper bound of the composition-dependent variation in
productivity is a flat line. This means that, for this sampling
effect model of exploitative competition for a single resource, there is no combination of many species that is ever
more productive or that uses resources more completely
than a monoculture of the best competitor. Similarly, for the
concentration of unconsumed resources, a diagnostic feature of the sampling effect is that the lower bound of its
variation is a flat line (figure 2.1B). Another diagnostic feature of the sampling effect is that, despite the large number
of species initially randomly assigned to a high diversity plot,
a single species (the best competitor and thus most productive species) is predicted to win in competition and eventually drive all other species extinct. This latter point lays bare
the sampling aspect that gives this model its name (Tilman,
Lehman, and Thomson 1997). The number of species initially sampled impacts ecosystem functioning even though
the species do not coexist, and all but one goes locally extinct. This illustrates one important aspect of diversity—that
diversity can affect an ecosystem merely because greater di-

21

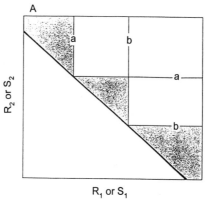

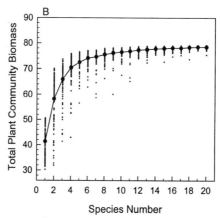

versity increases the chance that a superior species would be present in a locality.

Niche Differentiation Models

There are a large number of alternative models that predict long-term coexistence of many species. Each of these could be used to determine the effect of the number of coexisting species on ecosystem processes. Several have been so explored either analytically or in simulations (Tilman, Lehman, and Thomson 1997; Loreau 1998a; Tilman 1999b; Lehman and Tilman 2000). The models that predict long-term coexistence of species via interspecific differentiation in response to exploitative competition for environmental limiting factors we will call *niche differentiation models*. All such models explored to date make similar predictions, perhaps because they are based on simple mechanisms of exploitative resource competition, rather than on interference competition or multitrophic-level interactions.

Let us consider, first, a model of plants competing for two essential resources, such as nitrogen and phosphorus, or nitrogen and light (figure 2.2A). For many plant species to

---

FIGURE 2.2. A niche differentiation model of exploitative competition for two essential resources, based on the resource competition model of Tilman (1982). (A) Zero net growth isoclines are solid curves labeled *a* and *b*, for species *a* and *b*. Solid diagonal line is an interspecific trade-off curve, which shows the lowest levels to which any possible species could reduce levels of resources R_1 and R_2. Each point on this line is the corner of an isocline of a potential species. Shaded regions show levels of R_1 and R_2 that are not consumed by a community composed of species *a* and *b*. (B) For communities of randomly chosen species (i.e., randomly chosen points on the interspecific trade-off curve, with many such random communities per level of species diversity), simulations show that total community biomass is an increasing function of diversity. Each point is a result of a given simulation for a randomly chosen community of a given level of diversity, with the resulting competitive equilibrium shown for a spatially heterogeneous habitat. (C) Results for these same cases, but showing the levels of unconsumed resource. Figure modified from Tilman, Lehman, and Thomson (1997).

stably coexist when competing for two resources, it is necessary that they have interspecific trade-offs in their requirements for the two resources, with a species that is better at competing for one resource being worse at competing for another (Tilman 1982, 1988). It is also necessary, for multispecies coexistence on two resources, that the habitat be spatially heterogeneous in the supply rates of these two resources. In heterogeneous habitats, it is possible for a potentially unlimited number of species to stably coexist as long as they have interspecific competitive trade-offs. If we assume the existence of a pool of species with such trade-offs, we can then randomly assemble from it communities with different numbers of species, and determine how diversity impacts total community biomass (Tilman, Lehman, and Thomson 1997). For such models, community biomass is predicted to be an increasing function of diversity (figure 2.2B). Concentrations of unused resources are predicted to be a decreasing function of diversity (figure 2.2C). For biomass and resources, variance declines at higher diversity (see figure 2.2B, C). However, this niche differentiation model has a different signature than the sampling effect model. In particular, this model explicitly predicts that the upper bound of the composition-dependent variation in total community biomass is, itself, a monotonically increasing function of diversity, not the flat line of the sampling effect model. Thus there is no one-species plot that is as productive as some two-species plots, no two-species plots that are as productive as some three-species plots, and so forth. Similarly, the lower bounds of the resource curves are decreasing functions of diversity, and not flat lines (figure 2.2C). Finally, each randomly assembled community tends to maintain its initial diversity, rather than having a single species competitively displace all other species at equilibrium.

A second niche differentiation model makes qualitatively similar predictions (figure 2.3). In it, a habitat is assumed to have spatial or temporal heterogeneity with respect to two

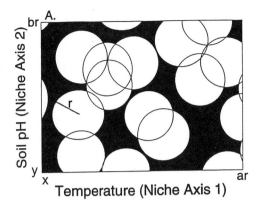

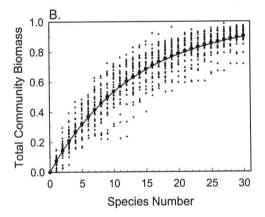

FIGURE 2.3. Niche differentiation in response to spatial heterogeneity in temperature and soil pH, modified from Tilman, Lehman, and Thomson (1997). (A) The physical conditions under which a species can survive are shown as a circle of radius r. The heterogeneous habitat has values of temperature that range from x to ar and of soil pH that range from y to br. (B) For randomly chosen species capable of living somewhere in this heterogeneous habitat, the amount of this habitat "covered" by viable species increases as species number increases, thus causing total community biomass to increase with diversity. Note that the number of species required to approximately saturate this relationship depends on the extent of habitat heterogeneity relative to the breadth (r) of conditions tolerated by species. A few species would totally cover a relatively homogeneous habitat, whereas a large number of species would be required to cover an extremely heterogeneous habitat.

limiting factors, such as soil pH and temperature, and each species is assumed to have a particular combination of values of these limiting factors at which it performs optimally. Temperature, for instance, could be considered to vary seasonally, with each species having a particular range of temperatures (and thus times during the growing season) for which it did well. Soil pH would vary spatially. For values of the limiting factors near the optimum of a species, this species would be the superior competitor. Other species would be superior competitors for other values. Each species would have a range of values, corresponding to a region in niche space, for which it did well. Each point on the plane of figure 2.3A would thus be the center of the niche of a species. Greater diversity would lead to greater coverage of the heterogeneous habitat, and thus to greater productivity (figure 2.3B). Tilman, Lehman, and Thomson (1997) offer an analytical solution to the case in which the niches are circles of radius r and the heterogeneous habitat is rectangular.

In a third niche model, the growth of all plant species is assumed to be constrained both by competition for a single limiting resource and by a physical factor, such as temperature (figure 2.4A). Each species is a superior competitor at a particular temperature, and performs progressively less well at temperatures further from its optimum. Temperature could vary seasonally (cool springs, hot summers, cool autumns) or from year to year. Temperature fluctuations allow multispecies coexistence, as in the models of Chesson (1986, 1989). This third model also predicts that ecosystem productivity would increase, on average, with diversity, and that the level of unused resource would decrease as diversity increases (figure 2.4B; Tilman 1999b). Note that the upper bound of the dependence of productivity on diversity is an increasing function of diversity and that the lower bound of the dependence of unused resource on diversity is a decreasing function of diversity, further suggesting that these

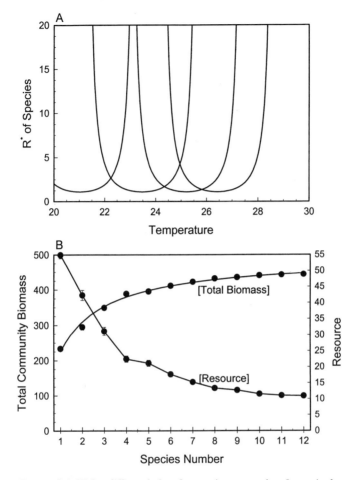

FIGURE 2.4. Niche differentiation for species competing for a single limiting resource in a habitat with spatial heterogeneity in temperature. (A) Curves show the temperature-dependence of competitive ability (lower R^* values lead to greater competitive ability) for species that differ in the temperature at which they have optimal competitive ability (their lowest R^* value). (B) For randomly assembled communities of such species, the average total community biomass increases and the average concentration of unconsumed resource decreases as diversity increases. In these simulations, most randomly chosen species persisted when competing. Figure modified from Tilman (1999b).

are signatures of niche differentiation models. Several other models of exploitative competition that predict the potential coexistence of numerous species also predict that average productivity (averaged across many randomly chosen communities of a given level of diversity) would increase with diversity and that the upper bound of this relationship would, itself, be an increasing function of diversity (Lehman and Tilman 2000).

In total, the niche models tested to date, which are models of simple exploitative competition in nonspatial habitats, predict that average productivity should be an increasing function of species diversity, but see chapters 9 to 11 for some exceptions to this trend. The sampling effect model of exploitative competition for a single resource and the niche models of exploitative competition differ in two major ways. Niche models predict the coexistence of species, whereas this sampling effect model predicts competitive dominance by one species and displacement of all other species at equilibrium. Secondly, the sampling effect model predicts that the upper bound of the dependence of productivity on diversity should be flat—with no higher-diversity plots ever being more productive than the most productive monoculture. In contrast, the niche models predict that the upper bound should, itself, be an increasing function of diversity, meaning that there would be no low-diversity community as productive as that of some higher-diversity species combinations. Loreau (1998b) suggested some alternative ways to distinguish between the sampling effect and alternative models.

Both niche and sampling effect models predict that differences in species composition cause variance in productivity within a given level of diversity. This variance can be large, especially at lower diversity, and indicates that compositional differences can have impacts on productivity or resource levels that are roughly of the same magnitude as those of diversity. Direct comparisons of the relative importance of diversity versus composition in such models are of limited utility because they depend on the range of diversity values considered, the range of species traits considered, and the size of the species pool.

DIVERSITY AND STABILITY

The diversity–stability hypothesis, as already discussed, has been a focus of interest and debate for many decades (e.g., Gardner and Ashby 1970; May 1972; McNaughton 1977; Pimm 1979, 1984; King and Pimm 1983; McNaughton 1993; Tilman and Downing 1994; Tilman 1996, 1999b; Doak et al. 1998). We will not review all of these studies here, but will briefly summarize our view of the current status of theory relating diversity to stability. In doing so, we will explore both the stability of populations within a community of competing species and the stability of the total community, that is, of the summed abundances of all competitors.

Measures of Stability

Part of the difficulty in the long-standing diversity—stability debate arises because there are many different definitions and meanings of stability in ecology (Pimm 1984). In deterministic mathematical systems, stability means the strength of return to equilibrium after perturbation—or more generally, the return to any prescribed behavior, such as to a limit cycle or a chaotic attractor, after perturbation. Therefore, in deterministic systems stability is commonly measured by eigenvalues (e.g., May 1972). Using this definition, May (1974) found that the number of species in a community of competing species influenced stability. In particular, he found a tendency for eigenvalue stability to decline as diversity increased.

In stochastic mathematical systems, the idea of stability is more related to long-term persistence of species and the extent of stochastic population oscillation relative to the average population size. The probability that a given set of parameters and initial conditions would lead to population

extinction can be precisely calculated for sufficiently simplified systems (e.g., May 1975), but in more complex
models, and indeed in nature, such calculations are difficult. Therefore, measures of stability commonly used for
stochastic systems, including natural systems, account for
fluctuations over time. Early contributors often thought of
stability in this way (e.g., MacArthur 1955; Elton 1958, pp.
146–50). The coefficient of variation—proportional to the
standard deviation divided by the mean—is one such measure often used (Tilman 1996; Doak et al. 1998; Tilman,
Lehman, and Bristow 1998). In this chapter we use a related
but slightly different stochastic measure of stability. As in
Tilman (1999b) and Lehman and Tilman (2000), we define
"temporal stability," S, to be the mean, μ, of a series of
values that vary over time, divided by the temporal standard
deviation, σ, in those values. This is a scale-invariant dimen-
sionless quantity similar to what is called the *signal-to-noise
ratio* in information theory. It is proportional to the reciprocal of the coefficient of variation, and is useful for values
like biomass that have positive means. It can be applied to
both near-equilibrium and to nonequilibrium systems (such
as systems in which diversity collapses to one or a few species
if stochastic fluctuations cease). Moreover, unlike eigenvalues, temporal stability is readily calculated from population sizes that have been observed in nature. The value of S
increases as stability increases. If mean size was constant but
temporal variation was to decline, temporal stability S could
increase without limit, because its denominator could approach zero. On the other hand, if temporal variations
progressively became larger relative to the mean, temporal
stability would decline.

Components of Temporal Stability

Temporal stability can be computed for any time-varying
values with a finite variance whose mean is greater than
zero. The S for the biomass or population density of an indi-

vidual species is simply the species' mean biomass or population density for a location divided by the temporal standard deviation in the biomass or population density for that location:

$$S_i = \frac{\mu_i}{\sigma_i}. \tag{2.3}$$

Here, S_i is the temporal stability of species i, μ_i is the mean abundance of species i, and σ_i is its temporal standard deviation. Similarly, for the entire community, temporal stability is the mean total biomass or abundance of all species in the community divided by the standard deviation in that total. For simplicity, we will refer to biomass, population density, or other measures of the abundances of organisms in a population or community simply as abundance.

These simple definitions can give considerable insight when their statistical and ecological bases are explicitly considered. The standard deviation of total community abundance, being the standard deviation of a sum of species abundances, includes both variance and covariance components, based on the well-known statistical relationship between the variances and covariances of variables and the variance in the sum of these variables. This means that the temporal stability of the entire community is therefore more complex than that of individual species. Specifically, the stability of total community biomass is

$$S_T = \frac{\mu_T}{\sigma_T} = \frac{\Sigma \text{ Individual species biomass}}{\sqrt{\Sigma \text{ Variance} + \Sigma \text{ Covariance}}} \tag{2.4}$$

Here, S_T is the temporal stability of the total community, μ_T is the mean community abundance, and σ_T is the temporal standard deviation in community abundance. Clearly, mean community abundance is just the sum, over all species, of the mean abundances of the species. The temporal standard deviation for the total community depends on the variances and covariances in the abundances of the species in the community. In particular, for a community containing N species, the summed variance term in the denominator of equation 2.2 is the sum, over all species, of the temporal variances of these species. The summed covariances is the sum over all possible unordered pairs of species of the temporal covariance in their abundances. Put another way, the entire denominator is the square root of the sum of all terms in the full $N \times N$ covariance matrix for abundances of these N species. The principal diagonal of this covariance matrix gives the variances in species abundances, and the off-diagonal elements are the covariances among all possible pairs of species.

Equation 2.4 shows that community stability will increase with diversity if (1) total community biomass increases, or (2) the summed variances decrease, or (3) the summed covariances decrease, or (4) some combination of the above occurs. Only one is necessary—for example, stability could increase with diversity even though the summed variances and covariances increased, provided that the summed abundances increased more rapidly as diversity increased.

TOTAL BIOMASS

An increase in total community biomass with increased diversity (point 1 above) is called *overyielding* (Harper 1977; Naeem et al. 1995; Tilman, Wedin, and Knops 1996; Tilman et al. 1997; Hector 1998). If the size of the fluctuations in total community biomass did not change with diversity, overyielding alone would stabilize a community because those fluctuations would be smaller relative to the mean total community abundance. We call this the *overyielding effect*.

COVARIANCE

The covariance term in the denominator measures interactions among species. It tends to be negative for competition, positive for mutualism, and zero for species that do not interact at all. If the sum of all pairwise covariance terms becomes more negative as diversity increases, which might happen because of competition among these species, the total community abundance would be stabilized, all else being equal. We call this the *covariance effect*.

VARIANCE

Both statistical and ecological processes impact the summed variances in the denominator of equation 2.2. The summed variances may decrease with increasing diversity because of the *portfolio effect* (Doak et al. 1998; Tilman, Lehman, and Bristow 1998), a form of statistical averaging. The portfolio effect arises because the average or sum of many randomly varying entities is, under most circumstances (see Tilman, Lehman, and Bristow 1998), less variable than the average individual entity. In particular, if a single random variable is replaced by two independent random variables drawn from the same distribution, each new variable being half the magnitude of the original, then the mean value of the sum of the two new variables will be the same as the mean value of the original, but the variance in the sum will be precisely half the variance in the original. The process continues as more variables are drawn (i.e., as diversity increases), with greater diversity leading to progressively lower variance in the total.

In economic rather than mathematical terms, a fixed amount invested in two statistically independently varying stocks would have (approximately) half the variance as the same amount invested in a single stock. In ecological terms, Doak et al. (1998) applied this idea to ecosystems and made the intriguing suggestion that if a fixed total biomass is partitioned between any number of independently varying species, then more diverse communities would have less variation relative to their mean because of statistical averaging.

While this effect of variance is precisely true for mathematical random variables, and very close to true for independent stocks, it need not be true for variables of dynamical systems in general. Hence it may or may not be true for ecological models or natural ecosystems (Tilman, Lehman, and Bristow 1998). The critical issue is the way in which the variance in the abundance of each species scales with its mean abundance, and how this scaling is influenced by intraspecific and interspecific interactions. If variance, σ^2, scales as the mean abundance of a species, m, raised to the z power,

$$\sigma^2 = cm^z \qquad (2.5)$$

then the portfolio effect will cause more diverse communities to be more stable only if $1 < z \leq 2$ (Tilman, Lehman, and Bristow 1998; Tilman 1999b). In particular, assuming that covariances are all zero, the temporal stability of a community of N species (S_N) will differ from that of the average monoculture (S_1) according to the relation

$$\frac{S_N}{S_1} = N^{(z-1)/2} \qquad (2.6)$$

where N is species number and z is the scaling parameter (figure 2.5; Tilman 1999b). Thus, if there are no covariances, the portfolio effect can cause increasing diversity to stabilize communities, just as Doak et al. (1998) discovered. Because analyses of extensive data on both natural communities and models have found that the variance scales with exponents between 1 and 2 (e.g., Taylor and Woiwod 1980; Murdoch and Stewart-Oaten 1989; Tilman 1999b), it seems likely that the portfolio effect would cause community stability to increase as diversity increases.

Diversity and Temporal Stability in Multispecies Models

This discussion illustrates that overyielding, the portfolio effect, and covariance can all influence the dependence of

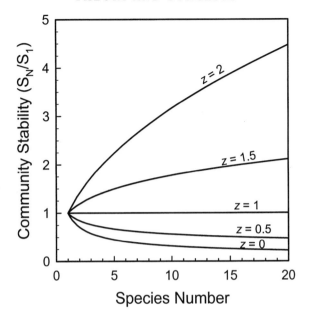

FIGURE 2.5. The stability of a community composed of N species relative to that of a community containing a single species, as it depends on species number (assuming covariances are all zero). This is analytically calculated for situations in which the scaling relationship, z, has different values. Note that community stability increases with diversity if $z > 1$. Figure modified from Tilman (1999b).

stability on diversity. However, the dependence of population and community temporal stability on diversity is influenced by the pattern of dependence of each of these factors on diversity. To better understand this dependence, it is necessary to consider how these parameters covary in models of multispecies interaction and coexistence. Let us consider one such model, the model of temporal niche partitioning for species that compete for the same limiting resource discussed earlier (figure 2.4; Tilman 1999b). By solving this model for many randomly assembled communities containing from 1 to 12 species, we were able to determine the dependence of the temporal stability of total community abundance and of species abundances on diversity. We found that community temporal stability was an almost linearly increasing function of the number of competing species (figure 2.5A). In contrast, the temporal stability of the average population was a declining function of diversity. Interestingly, the magnitude of the effect of diversity on community temporal stability was much greater than the magnitude of the effect of diversity on population temporal stability. Thus, much as reported by May (1972), we found a slight tendency for individual species to have their abundances destabilized as diversity increased. However, we also found a great tendency for the temporal stability of the entire community to increase as diversity increased.

On first inspection it may seem incongruous for individual species to become less stable as diversity increases and for their sum (which is community abundance) to become more stable. There is, however, no conflict between these two trends because the temporal stability of the entire community depends on the variability of individual species and on the interplay of the overyielding, portfolio, and covariance effects. Because these species are competitors, both the mean abundance of individual species and their temporal variance decline as diversity increases. Because population temporal stability is just the ratio of these two declining items, there is little effect of diversity on it. The slight decrease in population temporal stability with increased diversity occurs because variance does not decline quite as rapidly with diversity as does population abundance, especially at low diversity. Rather, the variance in the abundance of individual species scales with an exponent less than 1 when there are few competing species (Tilman 1999b).

The almost linear increase in community temporal stability with increased diversity results from the combined effects of overyielding, portfolio, and covariance effects (figure 2.6B). The effects of overyielding (figure 2.4B) are clear. S_T is the ratio of total community abundance to its temporal

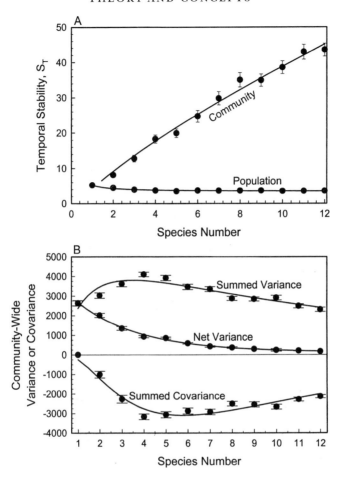

FIGURE 2.6. (A) Effects of diversity on population and community stability for a model (figure 2.4) of niche differentiation among species competing for a single resource in a spatially heterogeneous habitat. (B) Effects of diversity on the components (summed variance and summed covariance) of community-wide net variance. Modified from Tilman (1999b).

standard deviation. If nothing else were to change, increases in total community abundance would lead to greater temporal stability. The temporal standard deviation of total community abundance depends both on the summed variances and the summed covariances. The sum of summed variances and summed covariances is the net variance for the community. Net variance is a monotonically declining function of species number (figure 2.6B), which causes community temporal stability to increase as diversity increases. The summed covariances are always negative, as would be expected for a community of competitors (figur 2.5B). Importantly, summed covariances become increasingly more negative as diversity increases from one to a few species, but become progressively less negative for further increases in species number. This means that the covariance effect leads to increased temporal stability as species number increases only as species number increases from one to a few species. If nothing else were changing, further increases in diversity would cause community temporal stability to decline. However, summed variances show the opposite effect (figure 2.5B). Summed variances become somewhat greater (which is destabilizing) as species number increases from one to a few species, but decline beyond that point. This decline in summed variances increasingly stabilizes more diverse communities. The number of species at which the stabilizing effect of increased diversity switches from being caused by the covariance effect to being caused by the portfolio effect depends on the model considered, and on its formulation, but seems to range from about 4 to more than 20 species (Lehman and Tilman 2000). In total, the portfolio effect and the covariance effect, in combination with the overyielding effect, cause community temporal stability to be an almost linearly increasing function of species number.

We have found the same qualitative effects of diversity on community and population stability, and the same qualitative effects of diversity on overyielding, on summed variances, and on summed covariances in several other models (Lehman and Tilman 2000). One of these was a Lotka-Volterra model of multispecies competition quite similar to that used by May (1972) and the other was a modified broken stick model, which is a highly abstracted summary of resource competition. These predictions agree with the observed dependence of stability on diversity for a variety of systems studied by McNaughton (1993), for the effect of oak diversity on acorn woodpecker stability (Bock and Bock 1974, Koenig and Haydock 1999), and for the effects of grassland plant diversity on both the drought resistance of total community biomass and the stability of both individual plant species (slightly destabilized by increased diversity) and total community biomass (stabilized by increased plant diversity; Tilman 1996, 1999b). However, we do not yet have theories comparable to those summarized here that link diversity to population or community stability for cases other than communities of exploitative competitors. Such work needs to be expanded, both theoretically and experimentally, to predator-prey, mutualistic, and multitrophic-level systems, and to communities controlled either by antagonistic interactions or by an interplay of exploitative and interference interactions.

SUMMARY

In total, the theories reviewed here suggest that simple, well-known mechanisms of interspecific competition may cause community productivity and resource use and population and community temporal stability to depend on the number of species in a community. This demonstrates the plausibility of the long-standing suggestion that diversity can impact community and ecosystem processes. Moreover, different mechanisms linking diversity to ecosystem functioning, such as sampling effects versus niche differentiation effects, have different signatures, suggesting that these signatures can be used to evaluate the existence or relative importance of these mechanisms in laboratory or field experiments, or in comparative field studies. In all cases, the effects of diversity on community and ecosystem functioning result from simple, well-known ecological processes. In essence, diversity is a measure of the range of traits that occur in a community. For species that have traits that allow them to coexist when exploiting and competing for limiting resources, the same competitive mechanisms that allow coexistence also seem to cause greater diversity to lead to greater productivity, greater use of limiting resources, and greater stability. This, though, does not mean that greater diversity must always lead to these responses. Indeed, it is easy to hypothesize mechanisms of interference competition in which progressively better competitors have progressively lower carrying capacities, which could cause community productivity to decrease as diversity increases, and might also impact community temporal stability. Ultimately, we need to explore a range of mechanisms that might explain persistence of high-diversity communities, and determine, for each, what that mechanism implies for various aspects of ecosystem functioning. We then must learn which of these mechanisms operate in a given community, and the preponderance of these mechanisms in natural ecosystems, before we can offer a better generalization of the importance of diversity as a controller of ecosystem processes in nature.

All of these models linking diversity to ecosystem functioning have demonstrated that ecosystem functioning also simultaneously depends on community composition. Indeed, for the models presented here, about as much variance in ecosystem functioning is explained by diversity as is explained by the compositional differences among randomly assembled communities. Because it is well known that ecosystem functioning also depends on disturbance frequency and history, on productivity, on climate, and other variables, care must be taken in discussing the potential effects of changes in diversity. The results presented here suggest that the loss of species from a community of coexisting exploitative competitors would tend to decrease its productivity, decrease the extent of resource use, and decrease community temporal stability. However, the actual effect would depend on which species were lost, and on any concomitant shifts in other factors. Although it is now clear that diversity may influence ecosystem functioning, it is not yet clear how important diversity is relative to other factors nor how this depends on the status of the system of interest.

ACKNOWLEDGMENTS

We thank Nancy Larson for assistance and Ann Kinzig and Steve Pacala for comments on this paper. Supported by NSF/BSR 9411972 and the Andrew Mellon Foundation.

CHAPTER THREE

Experimental and Observational Studies of Diversity, Productivity, and Stability

David Tilman, Johannes Knops, David Wedin, and Peter Reich

From the early ideas of Darwin (1859), Elton (1958), Odum (1953), and others, to the work of May (1972), McNaughton (1977), Pimm (1979, 1984), King and Pimm (1983), to the more recent work of Schulze and Mooney (1993), Vitousek and Hooper (1993), Lawton and Brown (1993), McNaughton (1993), Huston (1997), Tilman, Lehman, and Thomson (1997), Loreau (1994, 1998a, 1998b), Ives, Gross, and Klug (1999), Tilman (1999b) and others, many alternative concepts and theories have been proposed linking the rate and direction of ecosystem processes to diversity. In this chapter, we evaluate these hypotheses by comparing their predictions to results of experimental and field studies. We first summarize experimental and field studies, and then present detailed analyses of recent, previously unpublished results of a biodiversity experiment that we have been performing in Minnesota since 1994 (Tilman, Wedin, and Knops 1996). We begin by evaluating field and laboratory evidence for the possible effects of diversity on stability, and then discuss evidence for the possible effects of diversity on productivity and nutrient dynamics.

DIVERSITY AND STABILITY

Early field evidence suggested that more diverse ecosystems were more stable in the sense of being less oscillatory and less susceptible to invasion by exotic species (Elton 1958). This, though, was seemingly contradicted by theoretical predictions based on several different types of models (Gardner and Ashby 1970; May 1972). Moreover, as pointed out in a review of more than 200 papers by Goodman (1975), the field evidence supporting the diversity–stability hypothesis was more anecdotal and observational than quantitative or experimental. Goodman (1975) concluded that the preponderance of evidence failed to support the hypothesis. McNaughton (1977) countered that the only strong contrary evidence came from theory, and accused the discipline of rejecting a hypothesis based purely on theory rather than on field evidence. In defending the diversity–stability hypothesis, McNaughton (1977) cited several observations and experiments in which greater plant community diversity was associated with greater resistance of a community trait to a perturbation. Despite this empirical work, and despite further developments of concepts and theory relating diversity to stability (e.g., Pimm 1979, 1984; King and Pimm 1983), the diversity–stability hypothesis was considered of little interest or relevance by many ecologists for the 15 years following Goodman (1975). However, Frank and McNaughton (1991) found, in a comparison of eight grasslands in Yellowstone National Park, that grasslands with greater plant species diversity had plant community compositions shift less in response to drought than those with lower diversity. Thus, at greater diversity, community composition was more resistant to drought. McNaughton (1993) evaluated 10 field studies to determine if there were an effect of diversity on community resilience or resistance to disturbance. He found evidence supporting the diversity–

43

stability hypothesis in 7 of the 10 cases. Although each case might be faulted for low replication or for lack of control for potentially confounding variables, McNaughton (1993) demonstrated that the preponderance of available evidence supported the diversity–stability hypothesis.

Although it was not gathered to evaluate the effects of diversity on community stability, the long-term monitoring of the population densities of hundreds of insect species by Taylor and Woiwod (1980) provides strong evidence suggestive of total community insect abundances being more stable at higher diversity. This occurs because of the statistical averaging effect of Doak et al. (1998). Specifically, Taylor and Woiwod (1980) found that the temporal variances in the abundances of individual species in these insect communities scaled as their abundance to a power of about 1.6, for which value the portfolio effect should hold (Tilman, Lehman, and Bristow 1998), with greater diversity leading to greater community stability. Further exploration of the data on these insect communities would be fruitful.

An intriguing effect of diversity on stability and productivity is provided by work on acorn woodpeckers (Bock and Bock 1974; Koenig and Haydock 1999). Acorn woodpeckers are highly dependent on acorns as a source of food, but oaks produce acorns as a mast seed crop. Masting means that there is great year-to-year variability in this resource. Bock and Bock (1974) and Koenig and Haydock (1999) discovered two interesting relationships. First, acorn woodpecker densities were greater in habitats with a greater number of species of oak trees. Second, there was a marked decrease in the year-to-year variability (CV) of acorn woodpecker abundances for habitats with higher oak diversity in California. Indeed, CV's fell more than threefold as the number of oak tree species increased from one to seven. Thus, greater oak diversity was associated with greater stability of acorn woodpecker populations. Although observational studies such as this are always subject to alternative interpretations, these studies suggest that the diversity of substitutable biotic resources available to a consumer species may influence both the population stability and the population density of the consumer species. This intuitively appealing possibility deserves further exploration.

A long-term nitrogen fertilization experiment at Cedar Creek Natural History Area in Minnesota provides evidence suggesting that greater plant diversity may lead to greater stability, here measured as resistance of community biomass to a major drought, the drought of 1987–88 (Tilman and Downing 1994; Tilman 1996, 1999b). In this study, plant diversity and composition were confounded because both were modified in response to various rates of nitrogen addition, but the large data set (207 plots), long time series (annual data from 1982 through the present), and the measurement of more than 20 potentially confounding variables allowed analyses that tested for effects of diversity while controlling for alternative mechanisms. A series of alternative analyses have all shown that resistance of total community biomass to drought was progressively greater at higher diversity (figure 3.1A), even when there was simultaneous control for numerous confounding variables (e.g., species composition, rate of nitrogen addition, productivity, root biomass, root:shoot ratios, functional group composition). Other factors, including species composition, were also significant determinants of community stability.

Analyses of data from these same 207 plots, but for non-drought years, showed considerable year-to-year variability in total community plant biomass. Interestingly, this year-to-year variability (CV) was significantly greater in lower-diversity plots (Fig. 3.1B), and this effect remained highly significant when there was statistical control for confounding variables (Tilman 1996). Composition was also a significant determinant of the degree of year-to-year variability in total community biomass. Thus, both in response to an extreme perturbation, a once-in-50-year drought, and in response to more

45

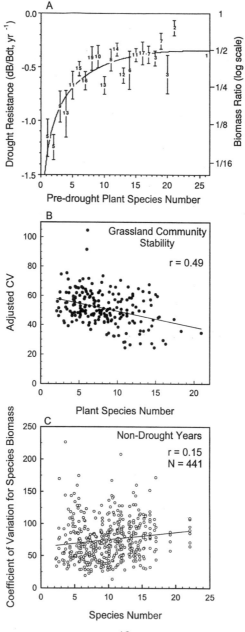

normal year-to-year variation in weather, greater diversity led to greater community stability, supporting the diversity–stability hypothesis (Tilman 1996, 1999b). However, analyses of the temporal stability of the populations of individual species showed that these had a slight but detectable trend toward being less stable at higher diversity (figure 3.1C, Tilman 1996). This latter result supports the prediction of May (1974) that increased diversity would lead to decreased population temporal stability ("if we concentrate on any one particular species our impression will be one of flux and hazard . . ."), and the greater stability of total community biomass supports his suggestion that increased diversity would stabilize community properties (". . . if we concentrate on total community properties such as biomass in a given trophic level our impression will be of pattern and steadiness"). In total, the results from a long-term study of plant community dynamics in these four Cedar Creek grassland fields support the predictions of the theory summarized in chapter 2. The theoretical predictions (May 1972, 1974; Tilman 1999b; Lehman and Tilman 2000) that increased diversity can destabilize population dynamics at the same time that it stabilizes community properties, and the

---

FIGURE 3.1. (A) Diversity and stability, measured as resistance to drought, in Minnesota grasslands, from Tilman and Downing (1994). Mean and standard errors are shown for each level of plant species number, with the number of plots having a level of diversity indicated by the numeral shown. Note that plots with about 15 or more species had aboveground biomass fall to about one-half of the predrought level, whereas those with 1 or 2 species had it fall to about one-tenth or one-twelfth. Analyses controlling for potentially confounding variables (Tilman and Downing 1994; Tilman 1996) still found highly significant effects on stability attributable to diversity. (B) CV's based on year-to-year variability in total community plant biomass, for this same experiment, show a significant tendency for plots with greater diversity to be less variable (from Tilman 1996, 1999b). (C) However, CV's for individual species show a slight tendency in the opposite direction (from Tilman 1996).

47

field evidence supporting this possibility, may help resolve the long-standing diversity–stability debate.

Although the Cedar Creek fertilization study supported the diversity–stability hypothesis, it lacked direct experimental control of plant diversity, and thus is subject to alternative interpretation (e.g., Givnish 1994; Huston 1997). This problem was overcome in three different experimental studies in which diversity was directly controlled. In a laboratory study of the effects of diversity in microbial communities, McGrady-Steed, Harris, and Morin (1997) found that the temporal variability of aquatic microcosms was significantly lower at higher diversity. Indeed, a fourfold increase in diversity led to about a threefold decrease in the temporal standard deviation in the rate of whole-community net respiration, a measure of ecosystem activity. The rate of microbial decomposition of particulate organic matter increased significantly with diversity in their study, supporting the diversity–productivity hypothesis for this decomposer community. They also determined the effect of diversity on susceptibility to invasion by a novel species by adding a few individuals of the species to all microcosms after the first experiment was completed. They found that greater diversity led to lower susceptibility to invasion, but that invader success was highly dependent on the composition of the resident community.

In another experiment, Naeem and Li (1997) created replicated laboratory microbial microcosms in which they experimentally varied both the number of species in various functional groups and the number of functional groups. They found that the variability in total community biomass among replicates was significantly lower when there were more species per functional group. This lower variance among replicates indicated that greater diversity led to greater reliability (predictability). Inspection of results of a greenhouse experiment (Naeem et al. 1995) also shows that greater diversity led to greater reliability in the total community biomass experiment (Tilman 1997b). Such an effect is analytically predicted for the sampling effect model (Tilman, Lehman, and Thomson 1997). However, for the niche models of exploitative competition that have been studied to date, the variance among replicates at a given level of diversity may monotonically decline as diversity increases, or it may reach a peak value at moderate diversity and then decline as diversity increases (Tilman, Lehman, and Thomson 1997; Lehman and Tilman 2000).

In total, these studies demonstrate that communities with greater diversity tend to be more stable, even though individual species in such communities may tend to be less stable. This agreement of theory, experiment, and observation lends support to the modified version of the diversity–stability hypothesis, first articulated by Robert May (1974), that states that increased diversity tends to stabilize community properties but tends to destabilize population properties. Moreover, this same body of work simultaneously demonstrates that species composition or functional group composition may be as important a determinant of stability as diversity. The next steps in understanding how diversity affects stability are to better map out the types of communities within which such diversity–stability processes occur or do not occur, to explore effects of diversity on other types of stability, and to find the conceptual and theoretical similarities and differences among eigenvalue stability, resistance, resilience, reliability, and other types of stability.

DIVERSITY, PRODUCTIVITY, AND NUTRIENT DYNAMICS

As brought to modern attention by McNaughton (1993), Darwin (1859) suggested that greater diversity should lead to greater productivity. McNaughton (1993), and Swift and Anderson (1993), evaluated this diversity–productivity hypothesis using then-available information. They found evidence suggesting that plant communities that contained more plant species were often more productive. Harper's (1977) summary of numerous studies of competition among grassland species also generally supports the diversity–productivity hypothesis because he found that mixtures of two-plant species that coexisted had greater biomass (over-yielded) compared to monocultures. Ewel, Mazzarino, and Berish (1991) found, in a field study in Costa Rica, that successional communities planted to many species generally retained soil fertility better than those planted to monoculture. As noted by Vitousek and Hooper (1993), this provided the first evidence suggestive of the diversity–nutrient retention hypothesis.

A direct test of the diversity–productivity hypothesis came from the greenhouse experiment of Naeem et al. (1995, 1996). By growing various randomly chosen combinations of 16 plant species 1, 2, 4, 8, or 16 at a time in small pots in a greenhouse, they found that mean total community biomass was a significantly increasing function of plant diversity (figure 3.2A). Inspection of the results of this short-term study may seem to suggest that they were not caused by resource-based niche differentiation (see chapter 2) but rather by some sort of a sampling effect, because the upper bound was not an increasing function of diversity but rather was relatively flat, or even declining (but also see Kinzig and Pacala, chapter 9). Interestingly, there were no higher-diversity pots that were as productive as the most productive monoculture, and only one higher-diversity pot was as productive as the next two most productive monoculture pots—again suggestive of a sampling effect. However, it is difficult to judge the importance of the most productive pots being monocultures because there were many more replicates of monocultures than of other diversity levels. The applicability of the theory of chapter 2 to this experiment is also difficult to judge because of the short duration of the experiment and the likelihood that its results are transients, perhaps driven by different maximal growth rates (Huston

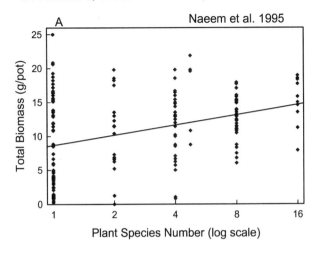

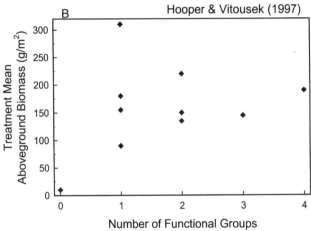

FIGURE 3.2. (A) Total plant community biomass for a pool of 16 plant species grown alone or in various randomly chosen combinations of 2, 4, 8, or 16 species in pots in a greenhouse experiment (Naeem et al. 1995; data for graph provided by Shahid Naeem, whom we thank). (B) Treatment means for a field experiment by Hooper and Vitousek (1997) in which California grassland species of four different functional groups were grown 1, 2, 3, or 4 functional groups at a time.

1997; Pacala and Tilman, chapter 7). A second experiment, performed in the Ecotron using many of these same species, had results suggestive of diversity causing higher productivity (Naeem et al. 1994), but its interpretation is clouded by nonrandom choice of species (Huston 1997; Hodgson et al. 1998). It was, though, an early study that helped highlight this area of inquiry experimentally and boldly attacked the as yet unresolved issue of the effect of the diversity of whole food webs rather than of a single trophic level (Lawton et al. 1998).

Hooper and Vitousek (1997, 1998) performed a field experiment, planted in 1992, in which they controlled plant functional group diversity and composition using plants common to California grasslands. They found, in the second field season, that functional group composition had a much greater effect on total community biomass than functional group diversity. Because their results (figure 3.2B; Hooper and Vitousek 1997) show that some plots containing one functional group had greater aboveground plant biomass than any of the plots containing two, three, or four functional groups, the results do not support the predictions of models of niche differentiation for exploitative competition. They did find that the utilization of soil nutrients increased significantly as diversity increased (Hooper and Vitousek 1998), but they did not publish raw data points, making it impossible to determine if the lower bound of variation in soil nutrient concentrations was flat or was a decreasing function of diversity.

In a four-month greenhouse experiment, Symstad et al. (1998) found that total plant biomass was significantly higher at higher diversity, and that most of this effect was attributable to the presence of legumes. They also determined the effects of the deletion of individual species on total biomass and found that the strength and direction of these effects depended on which species were present and which was deleted.

NEW RESULTS FROM THE CEDAR CREEK
BIODIVERSITY EXPERIMENT

A long-term field experiment, which we call Biodiversity I, provides a long-term test of the effects of plant diversity on total plant community biomass and nutrient dynamics. Tilman, Wedin, and Knops (1996) reported results from the second field season (1995). Here we present new results through the 1998 field season, which is the fifth field season of growth.

Methods

Biodiversity I was initiated by removing existing vegetation from an old field in 1993. It was planted in spring of 1994, and plots have been sampled annually beginning in 1995. The experiment contains 147 plots, each 3 m × 3 m, that were planted with 1, 2, 4, 6, 8, 12, or 24 plant species randomly and independently chosen from a set of 24 prairie-grassland species (Tilman, Wedin, and Knops 1996). There are 20 to 24 replicates of each level of plant diversity, with each replicate having its species composition randomly determined. Plots are manually weeded three to five times each growing season to remove any species not planted. The planted species are allowed to interact freely. All plots are watered weekly throughout the growing season. Plots are sampled annually for plant cover (visually estimated, to species), for the proportion of incident light intercepted by vegetation, and for extractable soil nitrate in the rooting zone (0–20 cm depths). Plots are occasionally sampled for soil nitrate below the rooting zone (40–60 cm depths), for aboveground living plant biomass via 10 cm × 100 cm clipped strips, and belowground plant biomass (root mass). More detailed methods are in Tilman, Wedin, and Knops (1996).

Soil Nitrate

We measured concentrations of extractable soil nitrate in 1995, 1996, 1997, and 1998. Nitrate concentrations both in the root zone and below the rooting zone were significantly decreasing functions of plant diversity (figure 3.3) every year. The lower bound of this variation was also a somewhat decreasing function of diversity in 1995 and was clearly a decreasing function of diversity by 1998 (figure 3.3B). The shape of this lower bound is consistent with predictions of resource-based niche differentiation models.

Community Cover and Biomass

We found in 1995, the second year, that total community plant cover (a nondestructive estimate of biomass) increased highly significantly with plant diversity (Tilman, Wedin, and Knops 1996). In 1995 the upper bound of the variation in total plant cover seemed to be approximately flat, with only three higher-diversity plots having greater total cover than the highest monoculture (figure 3.4A). On the surface, these 1995 plant cover results seem more consistent with the sampling effect model of competition for one resource than with resource-based niche differentiation models. As suggested by Huston (1997), one of the causes of the sampling effect in 1995 was the ability of *Rudbeckia serotina*, a rapidly growing species, to fill in open sites in plots. It reached its peak abundance in 1995, but fell to about one-tenth this abundance in subsequent years (figure 3.5A). In essence, it was a "Band-Aid" species that was capable of quickly filling in open sites, but that was displaced as slower growing superior competitors, such as *Schizachyrium, Andropogon,* and *Bouteloua*, increased in abundance (figure 3.5A). As shown in Pacala and Tilman (chapter 7), rapidly growing species can cause a transient state in which the upper bound in total community biomass is approximately flat, a point raised earlier by Huston (1997). By 1998, when

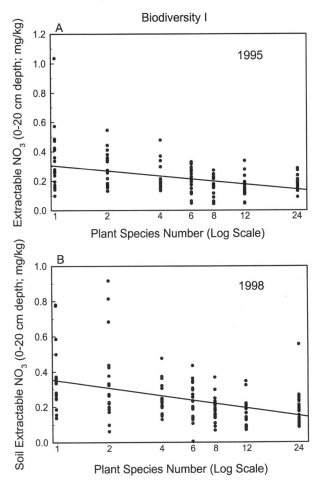

FIGURE 3.3. Extractable soil nitrate concentrations in the 147 plots of Biodiversity I in 1995 (A) and in 1998 (B).

total plant biomass (above- plus belowground biomass) was measured, the upper bound in its variation was clearly an increasing function of diversity, with 14 of the higher-diversity plots having greater total biomass than the highest monoculture (figure 3.4B). The plot with greatest total bio-

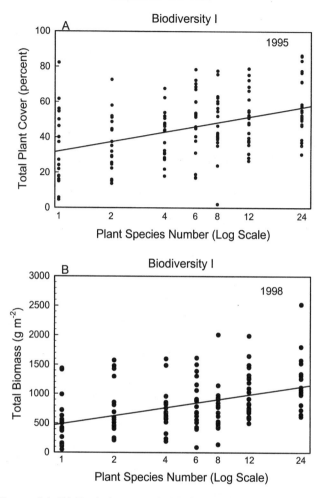

FIGURE 3.4. (A) Total plant cover in Biodiversity I in 1995. (B) Total plant biomass (aboveground plus belowground biomass) for Biodiversity I in 1998.

mass in 1998 was a 24-species plot that had 65% greater total biomass than the top monoculture. Thus, by 1998, the fifth field season of growth, the pattern of total biomass variation among replicates at different diversity levels reveals strong support for niche differentiation, as do the soil ni-

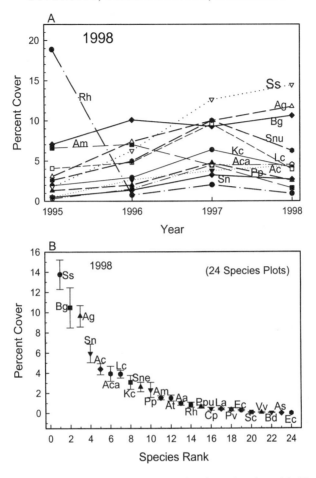

FIGURE 3.5. (A) Average percent cover of each species planted in Bio-diversity I for the highest-diversity treatment (24 species) from 1995 through 1998. (B) Dominance-diversity curve for the highest-diversity treatment of Biodiversity I in 1998, the fifth year of the experiment. Mean and standard error are shown for each species.

trate data (figure 3.3B). The importance of niche differen-tiation as a major cause of the greater community biomass of higher-diversity plots is further supported by the observed coexistence of the planted species (figure 3.5A). Contrary to the sampling effect model, none of the higher-diversity treatments in this experiment became dominated, at any time, by a single species that competitively displaced the other species. Rather, even after six years of growth, 22 of the 24 planted species persisted in the high-diversity plots, which is inconsistent with the predictions of the sampling effect model but supports niche differentiation models. These 24 species, when growing together, have attained relative abundances (figure 3.5B) much like their abundances in nearby grasslands. In particular, three of the four most abundant species in the 24-species plots are the native grasses *Schizachyrium scoparium, Andropogon gerardi,* and *Sorghastrum nutans,* which are the dominant species of nearby native grasslands. The fourth, *Bouteloua gracilis,* is a dominant of shortgrass prairie.

Another pattern lends further support to the role of niche differentiation in Biodiversity I. A comparison in 1998 of the abundances of each of these 24 species when growing either in monoculture or in two-species plots to their abundances when growing in the 24-species plots shows that these two abundances are positively correlated (figure 3.6). This shows that the species that perform best in monoculture or in two-species plots also tend to be the species that perform better at high diversity. However, the presence of other species greatly reduced the abundances of all species, including those that performed best in monoculture. The upper bound of the relationship of figure 3.6 represents all species being at least fourfold less abundant in the 24-species plots than in monoculture or two-species plots. The lower bound shows that some species that were highly abundant in monoculture, such as *Panicum virgatum,* were more than 40-fold less abundant in high-diversity plots. Thus, contrary to the sampling effect model, by 1998 the high biomass of high-diversity plots was not caused by the presence and dominance of those species that were most abundant at low diversity. Rather, much of the biomass of high-diversity plots was contributed by other species that coexisted with

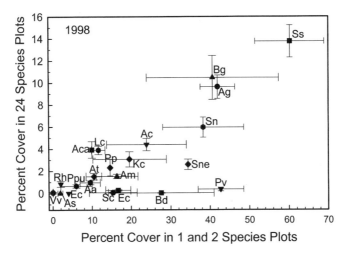

FIGURE 3.6. The relationship between the abundance (percent cover) of the species of Biodiversity I in low-diversity plots (1- and 2-species treatments) versus the highest-diversity plots, for 1998, the fifth year of the experiment.

these dominants. These coexisting species suppressed the biomass of the dominants well below that attained in low-diversity plots. In doing so, they collectively attained greater biomass than they collectively suppressed, causing some high-diversity plots to have greater biomass than any low-diversity plots (figure 3.4B). Such coexistence and overyielding is the signature of resource-based niche differentiation.

Our results suggest that sampling effects predominated initially in the experiment, but that the longer-term functioning of these communities was driven more by resource-based niche differentiation. During the initial stages of establishment, the observed responses would be driven more by differences in growth rates and in rates of vegetative spread (Pacala and Tilman, chapter 7). Such differences would cause sampling-like effects. However, as the habitat became filled in (especially belowground, for these low-nitrogen soils) interspecific competition would favor those species that were more efficient resource users, and allow those species to coexist that were differentiated in their resource use. Such differentiation is suggested by the shifts in species abundances that occurred during the experiment (figure 3.5A) and by the resulting dominance-diversity curve (figure 3.5B). Moreover, as discussed next, it is evident in the increasing clarity of the effects of treatments on the measured responses.

Species Number and Composition

Our results also can be used to determine the joint effects of diversity (number of plant species) and functional group composition (which plant functional groups were present) on the measured response variables. To do this, we classified the 24 species used in the experiment into four functional groups: cool season (C3) grasses, warm season (C4) grasses, legumes, and other forbs. There are 15 different possible combinations of these four functional groups when taken 1, 2, 3, or 4 at a time (i.e., $2^4 - 1$). Each combination represents a unique functional group composition. All 15 of these functional group compositions occurred in Biodiversity I. By classifying each plot by both the number of plant species planted in it (a continuous variable) and by its functional group composition (a categorical variable), multiple regressions can be performed to determine the effects of both diversity and functional group composition on all of the measured response variables (Table 3.1). We present multiple regressions based on Type I sums of squares to test first for the effects of the species number treatment on responses, and then determine if there are significant additional effects attributable to composition. Type III sums of squares are inappropriate because of the unavoidably high correlation between species number and composition.

The multiple regressions show that treatment effects have become much stronger through time, clearly demonstrating that the results observed in the fifth year of this experiment

TABLE 3.1. GLM Analyses of the Effects of the Number of Planted Species and of Functional Group Composition on Various Responses Measured in the Biodiversity Experiment

Variable Analyzed	Year	Overall			Species Number Treatment		Functional Composition Treatment	
		R^2	F Value	P	F Value	P	F Value	P
Total plant cover	1995	0.22	2.46	0.0034	25.10	<0.0001	0.84	0.6262
Total plant cover	1996	0.26	3.14	0.0002	32.68	<0.0001	1.03	0.4300
Total plant cover	1997	0.24	2.79	0.0009	29.65	<0.0001	0.87	0.5888
Total plant cover	1998	0.47	7.82	<0.0001	82.06	<0.0001	2.52	0.0033
Total biomass	1996	0.16	1.61	0.0792	4.03	0.0466	1.44	0.1449
Total biomass	1997	0.18	1.92	0.0269	4.63	0.0333	1.72	0.0584
Total biomass	1998	0.32	4.10	<0.0001	41.36	0.0001	1.44	0.1446
Light interception	1995	0.25	2.88	0.0006	13.86	0.0003	2.09	0.0160
Light interception	1996	0.24	2.69	0.0013	15.09	0.0002	1.80	0.0446
Light interception	1997	0.19	2.00	0.0195	10.71	0.0014	1.38	0.1712
Light interception	1998	0.31	3.91	<0.0001	25.02	<0.0001	2.40	0.0052
NO_3 0–20 cm	1995	0.36	4.89	<0.0001	18.36	<0.0001	3.93	<0.0001
NO_3 0–20 cm	1996	0.36	5.00	<0.0001	9.51	0.0025	4.68	<0.0001
NO_3 0–20 cm	1997	0.48	8.15	<0.0001	62.33	<0.0001	4.28	<0.0001
NO_3 0–20 cm	1998	0.60	13.02	<0.0001	34.79	<0.0001	11.46	<0.0001
NO_3 40–60 cm	1995	0.46	7.50	<0.0001	13.62	0.0003	7.06	<0.0001
NO_3 40–60 cm	1996	0.41	6.16	<0.0001	12.74	0.0005	5.69	<0.0001
NO_3 40–60 cm	1997	0.28	3.37	<0.0001	8.88	0.0034	2.98	0.0006
NO_3 40–60 cm	1998	0.51	9.11	<0.0001	23.12	<0.0001	8.11	<0.0001

Note. The number of planted species is entered first (Type I sums of squares); the functional group composition is entered second (Type I sums of squares).

were not transients. For instance, the R^2 values for total plant cover and for total plant biomass doubled from 1996 to 1998, and those for soil nitrate and light interception also increased markedly from 1995 to 1998 (Table 3.1). By 1998, from 30% to 60% of the variance in the response variables was explained by diversity and composition, indicating that these two variables were highly important determinants of the functioning of these ecosystems. The effects of the diversity treatment were significant ($P < 0.05$) for all 19 tests, and functional group composition explained a significant additional part of the variance for 12 of the 19 cases (Table 3.1). These analyses show an interesting pattern. Diversity has relatively stronger effects on total community abundance (measured as total biomass, total cover, or light penetration), whereas functional group composition has relatively stronger effects on soil nitrate.

In total, these analyses show that the effects of diversity and composition on ecosystem processes have increased as these communities have developed and as species abundances have adjusted in response to interspecific interactions. Moreover, they show that species diversity has relatively stronger effects on total community abundance than does functional group composition, but that composition has relatively stronger effects on soil nitrate levels than diversity. This difference may be simply explained by the opposite effects of legumes versus C4 grasses on soil nitrate, but their similar effects on total biomass. Legumes, being nitrogen fixers, tend to increase concentrations of soil nitrate, whereas C4 grasses have high root biomass and litter with high C:N ratios, both of which tend to reduce soil nitrate levels. This difference causes composition to be a major determinant of soil nitrate. In contrast, C4 grasses are highly productive because of their high efficiency of capture and use of nitrogen, and legumes are highly productive because of their ability to fix nitrogen. This means that these functional groups have similar effects on total biomass, thus causing species diversity—a measure of the chance that C4 grasses and/or legumes are present—to be a more important determinant of total biomass than functional group composition.

These results, in total, are consistent with diversity, in the long term, impacting productivity and soil nutrient levels more through a niche differentiation process than through a sampling effect. However, growth-rate–driven sampling effects are a viable explanation for the results observed initially. We began Biodiversity II, a parallel experiment that has larger but unwatered plots, a somewhat different mix of species, and annual burning, at the same time as and adjacent to Biodiversity I (Tilman et al. 1997). Its results are similar to those of Biodiversity I, but generally have somewhat higher R^2 values and show even stronger support for niche differentiation processes (Tilman et al., 2001). For instance, in Biodiversity II the upper bound of plot-to-plot variation in total biomass in 1998, 1999, and 2000 was a clearly increasing function of diversity. The top monoculture had lower total biomass than the average 16-species plot, and 60% of the 16-species plots had greater total biomass than the best monoculture in 2000.

Weedy Invasion and Fungal Pathogens

Knops et al. (1999) recorded the number of nonplanted species (mainly nonnative and native weedy species) that invaded the Biodiversity I plots, and their biomass at the time when they were removed from the plots. Significantly fewer species invaded higher-diversity plots, and the total biomass of invading species was lower in higher-diversity plots (figure 3.7). Further analyses suggested that the effect of diversity on invasion rate mainly was caused by the effect of diversity on soil nitrate. Specifically, Knops et al. (1999) found that, in a multiple regression that included both diversity and soil nitrate, soil nitrate had the stronger effect on invasion rate. Knops et al. (1999) also reported that the incidence of species-specific fungal leaf diseases was greater on

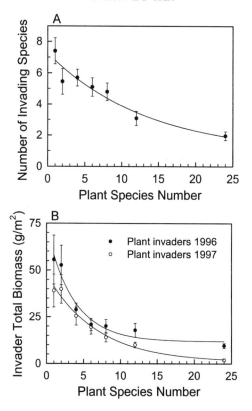

FIGURE 3.7. (A) Number of nonplanted species invading the various diversity treatments of Biodiversity I in 1996 and 1997, from Knops et al. (1999). These invading species are mainly native and nonnative annual plants and a few nonnative pasture grasses. Note that this graph does not include any of the 24 planted species that were part of the experiment, even if such species may have invaded plots into which they had not been planted. (B) Total biomass of all invading (nonplanted; not one of the 24 planted species) plant species in 1996 and 1997 for Biodiversity I. Figure modified from Knops et al. (1999).

average at lower diversity, with this effect seemingly attributable mainly to the effect of diversity on plant density. The lower density of each species in high-diversity plots seemed to decrease the incidence of their species-specific foliar fungal pathogens (Mitchell 2001).

Patterns in Native Grassland

We also explored the relationships between diversity, productivity and soil nitrate levels in a native, undisturbed grassland close to the two biodiversity experiments (Tilman, Wedin, and Knops 1996). We reported these results because they came from a single field, and thus represented a comparison of sites that were similar in many respects (same fire history, same parent soil, same source pool of species that were potential colonists of local sites). We found that total plant cover within this field generally was greater and soil nitrate generally was lower in more diverse plots (figure 3.8). This pattern is consistent with our experimental results and with the predictions of theory, but such correlational patterns must be viewed with caution because they could be confounded by other correlated variables. Loreau (1998a) used a model that linked environmental factors, biodiversity, and ecosystem functioning to explore this point. The model illustrated that correlational field data could be misinterpreted easily because of a confusion of cause-and-effect relationships.

Just such issues cloud the interpretation of the possible effects of island diversity on ecosystem processes for a study of 50 Swedish islands. In an intriguing study that showed links between island size, the frequency of wildfire, and community composition, Wardle et al. (1997) found that a suite of ecosystem traits also depended on island size, as did plant diversity. Because diversity, composition, and disturbance are confounded in this study, it is difficult to determine if any of the observed ecosystem responses are caused by diversity, or if they result from direct effects of fire frequency, or from compositional differences. Analyses of these data that statistically control for such collinearity are needed.

Because it is discussed in depth in another chapter in this book (Hector, chapter 4), we will not discuss here the effects of diversity and composition on the productivity of European grasslands. However, in its first two years, that exper-

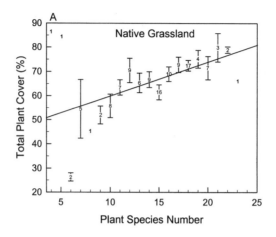

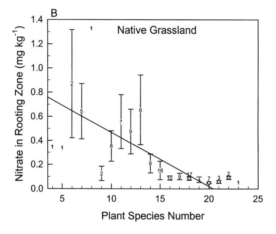

FIGURE 3.8. Correlations between plant species diversity in native savanna grassland openings and (A) total cover of the plant community and (B) extractable nitrate in the rooting zone, based on samples from 30 different locations in a single stand of native savanna in Minnesota (from Tilman, Wedin, and Knops 1996).

iment, replicated in eight sites across Europe, has provided results similar to the early results of the two Minnesota biodiversity experiments, thus suggesting that such patterns are generalizable.

SUMMARY AND SYNTHESIS

In total, the studies reviewed here have shown that diversity has significant effects on stability, productivity, and resource dynamics. In general, plant productivity is greater at greater diversity and this also corresponds with greater utilization of limiting soil resources. Short-term experiments showed weaker effects of diversity on productivity and soil nutrients than longer-term experiments. The shorter-term effects of diversity on ecosystem processes seem to be sampling effects, perhaps driven by different maximal rates of growth. If so, this would be consistent with the ideas of Huston (1997) and Grime (1997), and with Grime's (1979) view of plants becoming dominant by virtue of higher growth rates and resource preemption. In contrast, the longer-term effects of diversity seem to result from niche differentiation, likely driven by slower competitive shifts associated with interspecific differences in the efficiency of resource use and by differing patterns of spatial and temporal resource exploitation. Such interactions can still be causing changes in grassland species abundances after five or more years (e.g., Wedin and Tilman 1993).

A temporal transition from growth-rate-driven sampling effects to niche differentiation effects likely reflects the transient dynamics of competition. If species differ in their maximal growth rates, those plots that happened to have a fast-growing species (biased toward high-diversity plots by a sampling effect) would initially have greater total biomass. Under some circumstances this might even explain why some monocultures have greater biomass than any high-diversity plots in some short-term experiments (e.g., Naeem

67

et al. 1995, 1996; Hooper and Vitousek 1997; Pacala and Tilman, chapter 7). In contrast, niche differentiation effects would only become apparent after a longer competitive sorting of species.

A major difference between the resource-based sampling effect model (Tilman, Lehman, and Thomson 1997) and resource-based niche models suggests that we might expect a preponderance of sampling effects in greenhouse or growth chamber experiments and in modern high-intensity agriculture of annual crops, and of niche differentiation effects in field experiments and in relatively undisturbed natural ecosystems. This sampling effect model assumes that all plants compete in a spatially and temporally homogeneous habitat in which there is but a single limiting resource. Under such simple, homogeneous conditions, niche differentiation is unlikely to occur. Rather, a single species is likely to win in competition, and sampling effects should predominate. This may be what was happening in the greenhouse and growth chamber experiments (Naeem et al. 1994, 1995, 1996; Symstad et al. 1998), but they did not run for long enough to determine if coexistence or competitive displacement was the outcome of the competitive interactions. In contrast, in long-term field experiments in which there is spatial and temporal heterogeneity, and the potential for there to be several limiting resources, niche differentiation effects might eventually predominate, which they did in the two Minnesota grassland experiments (Tilman, Wedin, and Knops 1996; Tilman et al. 1997; Tilman et al. 2001). Moreover, the greater strength of niche differentiation effects in Biodiversity II compared to Biodiversity I is also consistent with this hypothesis. Biodiversity I is watered weekly, thus eliminating water as a potential limiting factor. Biodiversity II was watered initially, to help assure establishment of the plants, but was not watered after 1997.

Available theory that assumes that species mainly interact via exploitative competition for resources predicts that the productivity, nutrient utilization, and stability of communities should be greater at higher diversity (see review in chapter 2). A range of laboratory and field experiments support these predictions. Theory also predicts that species or functional group composition can be as important a determinant of ecosystem functioning as diversity, a prediction that also is supported by recent evidence. In all cases, the effects of diversity, as measured by species number, on community and ecosystem functioning are consistent with simple, well-known ecological processes. In essence, diversity is a measure of the range of traits that occur in a community. For species that have traits that allow them to coexist when exploiting and competing for limiting resources, the same competitive mechanisms that allow coexistence also seem to cause greater diversity to lead to greater productivity, greater use of limiting resources, and greater stability. This, though, does not mean that greater diversity must always lead to these responses. Indeed, it is easy to hypothesize mechanisms of interference competition in which progressively better competitors have progressively lower carrying capacities, which could cause community productivity to decrease as diversity increases, and might also impact community stability. Multispecies coexistence caused by a competition-colonization trade-off or by other sorts of successional niches can also lead to different patterns of the dependence of community productivity and perhaps stability on diversity (Kinzig and Pacala, chapter 9).

Ultimately, we need to explore a range of other mechanisms that might explain persistence of high-diversity communities, as is done in later chapters in this book, and to determine what each mechanism implies for various aspects of ecosystem functioning. We then must use field experiments to determine which, if any, of these mechanisms operates in various natural communities. The work on biodiversity to date demonstrates that diversity can have significant effects on many aspects of ecosystem functioning. However, we must know much more about the preponderance of various mechanisms across a wide range of natural ecosystems before we can offer rigorous generalizations about the relative importance of diversity versus other factors as controllers of ecosystem processes in nature.

ACKNOWLEDGMENTS

We thank Nancy Larson for assistance and Clarence Lehman and Ann Kinzig for comments on this paper. Supported by NSF/BSR 9411972 and the Andrew Mellon Foundation.

CHAPTER FOUR

Biodiversity and the Functioning of Grassland Ecosystems: Multi-Site Comparisons

*Andy Hector**

INTRODUCTION

An important goal of science is repeatability. This can prove difficult in ecology because environmental conditions such as the weather are never constant. Consequently, when faced with a small number of studies from a new area of research it is difficult to know how general the results are or whether they constitute special cases. When results conflict, does it reflect real differences in underlying biology, in environmental conditions, or in experimental design and methods? One way to progress is via robust meta-analysis of multiple studies (Osenberg, Sarnelle, and Goldberg 2000). Another is to conduct experiments with standardized methodologies that are replicated at multiple locations (Reader et al. 1994; Zak et al. 1994). In this chapter we report initial results from the BIODEPTH project, the first experiment to

*This chapter is based on research conducted by B. Schmid, C. Beierkuhnlein, M. C. Caldeira, M. Diemer, P. G. Dimitrakopoulos, J. A. Finn, H. Freitas, P. S. Giller, J. Good, R. Harris, P. Högberg, K. Huss-Danell, J. Joshi, A. Jumpponen, C. Körner, P. W. Leadley, M. Loreau, A. Minns, C.P.H. Mulder, G. O'Donovan, S. J. Otway, J. S. Pereira, A. Prinz, D. J. Read, M. Scherer-Lorenzen, E-D. Schulze, A-S. D. Siamantziouras, E. M. Spehn, A. C. Terry, A. Y. Troumbis, F. I. Woodward, S. Yachi, and J. H. Lawton.

71

manipulate diversity at multiple sites in a standardized way and to compare impacts on ecosystem functioning. We recently published a short account of the response of aboveground productivity (Hector et al. 1999) and a more extensive report is currently in preparation for publication elsewhere. Here we provide an overview of this work and discuss it in the broader context of other biodiversity research. Because the study of the impact of biodiversity on ecosystem functioning is a relatively new subject area of ecological research and because it has been extensively reviewed elsewhere (see Schläpfer and Schmid 1999 and Schmid, Joshi, and Schläpfer, chapter 6, this volume) we mainly confine our comparisons to the most similar experiments, that is, those that directly manipulate diversity by assembling communities of different diversity and monitoring changes in ecosystem processes.

THE BIODEPTH PROJECT

The BIODEPTH project (BIODiversity and Ecosystem Processes in Terrestrial Herbaceous systems: experimental manipulations of plant communities) investigated the impact of biodiversity on ecosystem processes by manipulating plant diversity directly in the field, establishing plant communities of planned diversity from seed. The general approach is similar to the biodiversity experiments of Tilman and colleagues at Cedar Creek (Tilman, Wedin, and Knops 1996; Tilman et al. 1997; Tilman 1999b) and Hooper and Vitousek in Californian serpentine grassland (Hooper and Vitousek 1997; Hooper 1998; Hooper and Vitousek 1998) and is described further in several other recent publications (Diemer et al. 1997; Lawton et al. 1998; Hector et al. 1999; Mulder et al. 1999; Scherer-Lorenzen 1999; Spehn et al. 2000a, b; Troumbis et al. 2000). The same experiment was conducted at eight different grassland fieldsites in Germany, Portugal, Switzerland, Greece, Ireland, Sweden and in the

United Kingdom at Sheffield and Silwood Park (near London). Communities of planned diversity were established from seed after removing the established vegetation and seedbank. The communities were maintained by regular weeding and managed as hay meadows. In total, the experiment comprised 480 experimental plant communities that varied in the numbers and types of species and plant functional groups (grasses, legumes, and other herbs) to compare the numerical, or richness, components of diversity with the compositional component. To do this, each level of species or functional group richness was represented by several replicate plant communities, each a different species or mixture of species. A pool of grassland species was identified at each fieldsite and plant communities assembled by random selection with some constraints (detailed in the references given above)—for example, most polycultures contained at least one species of grass (with the exception of some plots in Sweden and Ireland). Gradients of species richness with five levels were used at all sites, ranging from monocultures to a maximum diversity approximately equal to average background levels in plots of similar size in unmanipulated areas of grassland at each site. A secondary manipulation varied the number of plant functional groups independent of numbers of species. However, it is not possible to fully separate these two components across the full gradient of diversity (see below). This means that, in analyses, the effects of these two variables are partially correlated. In Hector et al. (1999) we present the order of analysis that best matches our a priori hypotheses that gives priority to species richness. In this chapter we present an overview of this and additional analyses of the data on aboveground biomass production during the second year of the experiment, comparing the relative importance of these two components of diversity, identifying their separate effects and to what degree they are correlated.

Multiple Influences on Productivity

One of the key features of the BIODEPTH experimental design is the ability to partition the results into those associated with location, with numerical components of diversity (i.e., species richness and functional group richness), and with the remaining effects of species composition. Between them, these variables explained 90% of the variation in an analysis of variance (table 3, Hector et al. 1999) of the productivity in our experimental plant communities, with the main effects of location, richness (numbers of species and functional groups combined), and composition accounting for about 30%, 20%, and 40%, respectively (figure 4.1). Location and the richness terms all had highly significant effects (table 3, Hector et al. 1999). Although accounting for the largest portion of variation in the experiment, species composition also has a large number of degrees of freedom associated with it and was not statistically significant. However, there was a significant interaction between location and species composition. This means that where the same species or mixture of species occurred at more than one site, the mixtures achieved significantly different yields, and that the realized productivity of plots at a site differed significantly depending on which particular species or mixture was present.

Differences between Locations

Productivity varied significantly between locations (Hector et al. 1999), with yields from northern and southern sites in Sweden, Portugal, and Greece about half those from more central European sites. Many factors that vary between localities combine to influence productivity, including soil type, number of harvests, and so on. Not surprisingly, climate appears to play a major role, with productivity in northern and southern locations probably frequently limited by temperature and water. With only eight fieldsites, we

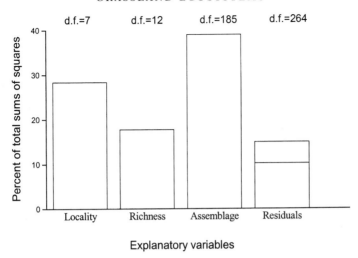

FIGURE 4.1. Multiple control of the productivity of the BIODEPTH plant assemblages. Location effects are differences between the eight fieldsites, richness is the combined effects of numbers of species and functional groups, assemblage effects are the remaining differences between the different species and mixtures of species, all expressed as percentages of the total sums of squares with degrees of freedom shown above. Residuals contain the interaction between composition and location (about 5%) and the overall remaining differences including nonsignificant block and interaction effects (10%).

are limited in how far we can explore differences between locations, given the large number of variables that vary. Nevertheless, differences between locations are useful for putting the size of biodiversity effects in context.

Species Richness versus Functional Groups

The BIODEPTH experiment is primarily designed to manipulate species richness. However, there is no magic effect of numbers of species per se. The potential mechanisms have been widely discussed (e.g., Naeem et al. 1994; Tilman, Wedin, and Knops 1996; Hooper 1998) and can be traced back as far as Darwin (McNaughton 1993): Any effect arises from biological differences between species and from species interactions that have consequences for ecosystem functioning. We examined these functional differences between species specifically with a secondary, less extensive manipulation of plant functional types. This manipulation was intended to test the hypothesis that it is through changes in functional diversity that species richness has effects, and to examine how far this functional diversity can be represented by differences among grasses, nitrogen-fixing legumes, and other herbaceous species.

As mentioned above, it is impossible to have a fully-orthogonal separation of the effects of species and functional group richness across the full gradient of diversity featured in our experiment (e.g., communities with one and two species can only have one, and one or two functional groups, respectively), causing the amount of variation associated with the different terms to be dependent on the order in which they appear in the statistical model. So far we have dealt with this issue in two ways (Hector et al. 1999).

First, we reported the combined effect of the two richness terms, which does not involve issues of order of fitting. Together the two richness components account for nearly 20% of the total variation. The interaction between species and functional group richness turns out to be nonsignificant, although our power in testing this effect may be limited due to the only partial separation of these variables. When this interaction is dropped from the model, the variation can be partitioned into the main effects of the two richness terms with two alternative sequences giving priority to one term or the other (that is, fitted first in the model or removed last with the backward selection procedure we adopted). The two richness terms turn out to have roughly similar-sized effects when given priority in the model, each explaining about 15% of the total variation. The significant main effects of these two variables *when each is given priority* are shown in figure 4.2.

The second approach was to present the sequential anal-

(A)

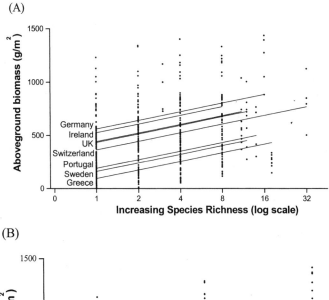

(B)

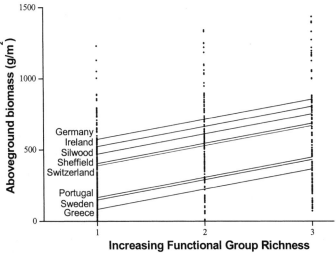

FIGURE 4.2. (A) Aboveground productivity declines with the loss of species richness. While average productivity and species richness relationships differ in detail at each location model, simplification by backward selection reveals an overall log-linear decline in aboveground biomass. Note that species richness is shown on an increasing scale for comparison with other chapters (not a decreasing scale as in Hector et al. 1999). (B) Aboveground productivity declines with the loss of plant functional groups.

77

ysis that matched our a priori hypotheses and experimental design, giving priority to our primary variable of interest, species richness. As a first approach for identifying the important effects in our complex design, we performed a sequential analysis with all of the data combined in one general linear model to simultaneously examine the relative effects of location, richness, and composition. The analysis is quite complex so we give an overview of our procedure here. We began using ANOVA (making fewer assumptions than selecting a regression model a priori) and used backward selection of terms from the maximal model (a very readable explanation of the general approach can be found in Crawley (1993)). There were no significant interactions, except the locality by assemblage term discussed above, so we ended up with a relatively simple model containing main effects of location and richness. Where we found significant effects in the initial ANOVA, we then attempted to simplify the model further by investigating whether it was possible to replace the ANOVA terms with simpler linear regressions without having significant scatter around them (in effect a linear regression followed by an ANOVA fitting species richness factor levels to encompass all the higher-order polynomial terms). We found we could simplify the analysis so that our final model contained significant log-linear effects of species richness and linear effects of functional group richness, with the regression lines having different elevations to reflect the different productivities at different sites. The log-linear effect suggests initially weak but accelerating effects of the loss of species. A log-linear model with diversity on a $\log_2$ scale predicts that each halving of the number of species reduced productivity by approximately 80 g/m^2, equivalent to reductions of 10–20%, depending on the background productivity of the different fieldsites (figure 4.2a).

The log-linear effect of species richness summarizes the overall average response, and this general outcome may provide a useful "rule of thumb" (Tilman 1999a) for the probable effect of the loss of species in the absence of more detailed information. However, in individual site analyses, the regression model best describing the species richness effect (based on the residual R^2) differed with some relationships stronger than the overall log-linear pattern, with others, weaker (figure 2, Hector et al. 1999). Notably, there was no effect of species richness at the Greek BIODEPTH site (but see Troumbis et al. 2000). One of our major tasks now is conducting more detailed analyses to try and identify how far this variation reflects ecological differences and how much is due to details of the specific experiments at the different locations. We discuss some of these differences below.

Exploring the main analysis further: In the model giving priority to species richness, functional groups also have a significant effect (Hector et al. 1999). This is consistent with our prediction that the effects of species on ecosystem processes will be related to the size of their functional differences. When the order of fitting is reversed the effects of species richness are not significant (table 4.1). Overall, the effects of the two variables are unavoidably largely correlated in our experiment, making it impossible to fully separate their effects. When considered in this way, about 12% of the variance is overlapping and attributed to whichever variable is given priority, with only around 3% uniquely attributable to both species richness or functional group richness after controlling for the effects of the other variable.

Richness versus Composition

Although the BIODEPTH experiment was primarily designed to examine the effects of numbers of species and functional groups it is also possible to examine other components of diversity. One approach is to ask how much of the compositional effects of the plant assemblages can be explained by individual species or groups (Hector et al.

79

TABLE 4.1. Summary of the Analysis of Second-Year Aboveground Biomass Giving Priority to the Effects of Functional Group Richness

Source of Variation	Degrees of Freedom	Sum of Squares	Mean Squares	F	P
Location	7	12,413,386	1,773,341	24.95	P < 0.001
Functional richness	2	6,353,715	3,176,858	34.63	P < 0.001
Linear contrast	*1*	*6,284,347*	*6,284,347*	*68.49*	*P < 0.001*
Deviation	*1*	*69,368*	*69,368*	*0.76*	*P = 0.3859*
Species numbers	10	1,415,958	141,596	1.54	P = 0.1292
Log-linear contrast	*1*	*169,471*	*169,471*	*1.85*	*P = 0.1761*
Deviation	*9*	*1,246,487*	*138,499*	*1.51*	*P = 0.1491*
Locality × F. richness	14	1,850,102	132,150	1.86	P = 0.0791
Locality × S. richness	20	1,372,882	68,644	0.97	P = 0.5239
Assemblage	152	13,946,043	91,750	1.29	P = 0.2171
Locality × Assemblage	28	1,990,165	71,077	3.74	P < 0.001
Residual	235	4,465,007	19,000		
Total	468	43,807,258	93,605		

Notes: Richness effects are partitioned into linear regressions and the deviation from linearity (in italics). The location and assemblage terms are tested against the locality-by-assemblage interaction, richness terms against the assemblage term, and locality-by-richness interactions against the locality-by-assemblage interaction. The alternative analysis giving priority to the effects of species richness is presented in Hector et al. 1999, where further details on similar analyses are available.

1999). When explanatory variables coding for the presence or absence of 71 of the most common species are added individually, 29 of them have significant effects. While many species had relatively small effects, some of the nitrogen fixers had a large influence as expected. These contrasts must be interpreted with some caution, as the experiment was not specifically designed to perform these multiple post-hoc tests. For example, by fitting each species individually, their effects may be exaggerated because many species will occur with each other in different mixtures and their effects are again likely to be correlated, just as the two richness effects were. This means that when attempting to test the effects of single species we may actually pick out the effects of several. Nevertheless, the results highlight the importance of the effects of many different individual species and show how they contribute significantly to the overall effect of diversity. Fitting similar variables, coding for the presence or absence of individual functional groups, reveals significant effects of legumes and herbs (figure 3, Hector et al. 1999). Our ability to test for the effects of grasses is limited since most polycultures contained them. This constraint should be taken into account when comparing our results with those of randomized designs with no such constraints (see Tilman et al., chapter 3, this volume). For example, if grasses are on average more productive than forbs this could cause average polyculture yields to be greater compared to monocultures than in fully randomized designs. We constrained our polycultures in this way to mimic natural grasslands (which, by definition, always contain grasses) with greater realism.

An alternative approach is to replace the richness terms with variables examining other components of diversity (again, the results must be interpreted with some caution since the experiment was not specifically designed to test these terms). The effect of functional group composition was examined by classifying communities into all the different possible combinations of grasses, legumes, and herbs (grasses, legumes, herbs, grasses plus herbs, . . . , grasses plus legumes and herbs). There was a highly significant effect of functional group composition that interacted with location ($F_{37,17} = 3.04$; $P < 0.01$), indicating the relationship was complex, varying between sites. Species richness within mixtures with different functional group compositions also had a significant effect ($F_{10,156} = 2.34$; $P < 0.05$).

Effects of Nitrogen Fixers

While experiments examining biodiversity and ecosystem functioning are a relatively recent area of ecological research there are, of course, many older studies from agriculture and ecology that are relevant. Trenbath (1974) still provides one of the best summaries of much of this work. One of the main conclusions from this and related reviews

(Harper 1977) is the well-known fact that the yields of grass crops can be increased by growing them in combination with legumes due to the atmospheric nitrogen they fix. Many of these earlier experiments were short-term, some took place under glasshouse conditions, and most involved only a small number of species, usually two. One of our aims is to compare these earlier results with those seen from our more diverse communities under more natural conditions and over the full duration of the experiment (we have three years of data from all of our sites, four or five years of data from some, and hope to extend this further). So far, analyses examining the effects of the presence of legumes in the experimental communities reveal a strong interaction with location ($F_{7,23} = 8.77$; $P < 0.001$). Legumes clearly have a big impact on productivity, but the strength of this effect varies between sites. The effects of legumes were particularly marked in Germany (Scherer-Lorenzen 1999) and Switzerland (Spehn et al. 2000b), probably due to the high availability of phosphorous. However, there was little effect of legumes at some sites at this stage of the experiment. After considering the effects of legumes, species richness still had a highly significant effect ($F_{10,144} = 3.48$; $P < 0.001$), which did not differ significantly between sites (locality-by-species richness interaction: $F_{22,23} = 1.5$; $P > 0.05$). Although the general effects of legumes on productivity are well known, they raise many interesting questions in our experiment. For example, why does the effect differ between locations? How does it change over time? What are the associated effects on other ecosystem processes, and how do all of these depend on the number and type of legume and nonlegume species present?

The Sampling Effect and Biodiversity Mechanisms

Biodiversity manipulation experiments have prompted much debate over their experimental design and interpretation. This is perhaps not surprising for a new area of research, particularly one that deals with a multifaceted subject such as biodiversity. Many of these issues have been widely discussed elsewhere (see Schmid, Joshi, and Schläpfer, chapter 4, and Tilman and Lehman, chapter 2, this volume), but two issues are of particular interest here. Both involve the sampling process that is involved when one randomly assembles communities from a pool of species (I will refer to them generally as 'sampling effect'). First, it has been suggested that positive relationships between productivity and increasing diversity may be explained by the greater chance of high-diversity assemblages containing species with high germination rates, leading to higher values for plant cover and biomass (Huston 1997). We have addressed the cover issue in two ways (Hector et al. 1999). First, we included cover as a covariate in the analyses of biomass. Second, we excluded plots from the analyses with less than 80% cover. In both cases, the highly significant effects of species and functional group richness on productivity remained in our analyses, suggesting that our biomass patterns cannot be explained solely by the sampling effect model as applied to plant establishment and cover (Hector et al. 1999).

In the second related sampling effect model, more diverse plant assemblages are assumed to have a higher probability of containing, and becoming dominated by, species with high biomass (Huston 1997; Tilman, Lehman, and Thomson 1997). Again, opinions differ on how to interpret this effect (Aarssen 1997; Tilman, Lehman, and Thomson 1997; Lawton et al. 1998; Symstad et al. 1998; Lawton 1999; Van der Heijden et al. 1999) so that it is clearly essential to identify and separate sampling effects from other biodiversity effects (Loreau 1998b). Several papers address methods for separating different biodiversity effects (Garnier et al. 1997; Hector 1998; Loreau 1998b), and some approaches have already been applied in previous experiments (Tilman et al. 1997; Wardle, Bonner, and Nicholson 1997; Hooper

1998; Hooper and Vitousek 1998). We performed several different tests of the ability of the sampling effect model to account for the productivity patterns observed in our experiment.

Testing the Sampling Effect

TEST 1: DO DOMINANT SPECIES MAINTAIN THEIR MONOCULTURE BIOMASS ACROSS THE DIVERSITY GRADIENT?

The sampling effect model states that the yield of a mixture of species should equal the biomass of the species in that mixture with the highest yield when grown in monoculture (Huston 1997; Tilman, Lehman, and Thomson 1997). We tested this assumption by looking for clear dominance of our plant assemblages (i.e., high relative abundance) using simple univariate regressions of the estimated seedling biomass (total biomass per species in a mixture divided by the reciprocal of its sowing density) of 14 of the more common species (those occurring in 35 plots or more) on diversity. It was not possible to count numbers of adult individuals as many species were clonal, so estimated values per seedling were used to gain measures of performance of the species initially sown in our experiments. Because the performance of species varied at different sites we took the residuals from models containing location and block terms, added them to the overall mean, and performed the regressions on these adjusted values. The assumption of the sampling effect model—that the productivity of a mixture of species is equal to the productivity of the dominant species in the mixture—implies that dominant species maintain their biomass across the diversity gradient (i.e., flat regression lines). To achieve this in our experiment in the short term, the larger dominant species would have to increase their biomass per individual seedling as diversity increased and their sowing density decreased (our experiment used a constant total density of 2,000 seeds per square meter divided equally among the number of species in each mixture). In contrast, the regression analyses revealed that 13 of the 14 species had significantly reduced biomass in higher-diversity plots relative to that in monoculture. Most of the species we analyzed did not maintain their monoculture yield across the diversity gradient. The one species whose biomass across the diversity gradient did not differ significantly from its monoculture yield was *Lolium perenne*, which was the fifth-highest-yielding species in monoculture overall. It was clear that, in general, the species that were most productive in monoculture did not maintain their biomass across the diversity gradient by strongly dominating mixtures as predicted by the sampling effect model.

The assumption of strong dominance by a single species made by the sampling effect model is fairly stringent. Perhaps communities are instead co-dominated by several species with relatively high biomass. This process clearly does operate at least at some of our sites. For example, in Sweden there was a positive correlation between monoculture biomass and relative abundance in mixture although this could not adequately explain the productivity patterns overall (Mulder et al. 1999). What if we relax the assumption of dominance by the single largest species and look at the performance of many individual species?

TEST 2: EVIDENCE FOR COMPLEMENTARITY BETWEEN
SPECIES? REGRESSION TESTS

The sampling effect model predicts that the dominance of some species across the diversity gradient should be compensated by decreases in the biomass of other subordinate species due to competition. On the other hand, niche complementarity models predict some reduction in competition due to processes such as resource partitioning. We used regression analyses of the estimated per-seedling biomass of individual species, similar to those reported above, to test for competitive effects (Hector et al. 1999). Consider three different possible results. First, if estimated biomass per plant of a given species does not vary across the diversity gradient it suggests that intraspecific competition, which is more common in low-diversity mixtures and monocultures, is equal to interspecific competition in higher-diversity mixtures. Second, if biomass per seedling of a given species declines with increasing diversity it suggests that it tends to lose in competition with plants of other species relative to plants of its own species. Third, if biomass per seedling of a given species increases with increasing diversity it suggests that it tends to find competition with its own species more intense than with other species. Fewer species than expected, only two of the 14, showed significant decreases in individual biomass with increasing numbers of species, suggesting some reduction in interspecific competition in more diverse communities (table 4, Hector et al. 1999). These results are similar to those of Tilman, Lehman, and Thomson (1997) who performed essentially the same test (but using total biomass per species rather than estimated biomass per seedling) and came to a similar conclusion.

TEST 3: EVIDENCE FOR COMPLEMENTARITY? TESTS USING RELATIVE YIELD TOTALS

The sampling effect model contains no complementary interactions between species. Therefore, increases in the dominance of some species due to competition should be balanced by decreases in the biomass of other subordinate species. Test 2 uses regression techniques and the total biomass of the more common species to examine this assumption. However, this assumption can also be tested for individual mixtures where all of the species in a mixture have also been grown in monoculture by using relative yield totals (RYTs) and related measures (Hector 1998). This need to have all species grown in monoculture means that this test will often only be applicable to a subset of the mixtures in a biodiversity experiment as most experiments will involve too many species for all of them to be grown alone. Care must be taken to ensure that this subset is representative of the overall pattern. For example, in our experiment we were only able to calculate RYTs for a few plots at our more species-rich sites in Germany, Switzerland, and Greece, and these were mainly low-diversity plots. The productivity response in Germany was linear and in Switzerland, log-linear, while in Greece there was no relationship. On balance the loss of the higher-diversity plots from these sites weakens the overall relationship between diversity and productivity but the general pattern remains qualitatively similar. It is important to consider this when extrapolating the results of these tests on subsets of the data to explain patterns in the main analysis. The relative yield (RY) of a species is simply its yield in mixture expressed as a proportion of its yield in monoculture, and the relative yield total of a mixture is simply the sum of the individual RYs. Species winning in competition will have large RYs and species losing in competition, small RYs. In the absence of complementarity, the increase in the RYs of some species will be exactly balanced by decreases in the RYs of others so that RYT = 1. The sampling effect model assumes an RY of approximately 1 for the dominant species in a mixture and that RYT cannot be greater than 1. We consider the RYT results below together with related tests for overyielding.

TEST 4: EVIDENCE FOR OVERYIELDING?

Test 3 uses RYT = 1 (or D_{bar} = 0, see Loreau (1998b)) as a new null hypothesis against which to test the sampling effect. Preliminary results reject this new null hypothesis and support the existence of complementary interactions in our experimental plant communities. However, while RYT > 1 is consistent with complementarity, it does not mean that the mixture necessarily outperforms the best of the component species when grown alone (Hector 1998). This more stringent situation, termed "overyielding" in the agriculture and plant ecology literature, requires a more conservative test. We tested for overyielding in our plant communities by identifying the dominant species in each mixture. Where a dominant species was also grown in monoculture we could then compare the performance of a mixture with the performance of the dominant species when grown alone using an index we call D_{max} (Loreau 1998b). The D_{max} index also standardizes for differences in absolute biomass of different species and mixtures. D_{max} is scaled so that a mixture that performs the same as a mono-culture of the dominant species in that mixture as predicted by the sampling effect model, has a D_{max} value of zero. Therefore, $D_{max} = 0$ provides another, more conservative null hypothesis against which the sampling hypothesis can be tested. One advantage of this test is that it does not re-quire all species to be grown in monoculture, only the dom-inants. This means that it is likely to be applicable to more of the mixtures in a biodiversity experiment. Nevertheless, we can only apply it to a subset of our plots. For the second-year harvest data from our experiment we could calculate D_{max} for 271 out of a possible 308 polycultures.

We have not yet completed comprehensive analyses of the relative yield or overyielding data. However, we can answer two general questions. First, are the data consistent with the predictions of the sampling effect? Second, is there evi-dence that complementarity increases with diversity? The grand mean RYT over all polycultures was 1.5 (SEM: 0.06), rejecting the sampling effect (T-test: $T = 25.27$, $P < 0.001$; $N = 204$) and consistent with complementarity in our plant communities. Similarly, the grand mean D_{max} over all poly-cultures of 0.35 (SEM: 0.061) was significantly greater than zero ($T = 5.69$; $P < 0.001$; $N = 273$), indicating that many mixtures performed better than even the best-performing species they contained. The results of ANOVA for both datasets were similar to the overall analysis of biomass. Rela-tive yields and overyielding differed between sites, and on average there was a general trend for overyielding to increase with diversity, although some low-diversity mixtures achieved high values and the trends were not simple progressive increases, indicating effects of composition.

SUMMARY OF THE BIODEPTH RESULTS

Overall, our results reveal multiple control of the productivity of our experimental plant assemblages by geographic location and the numbers and types of plant species and functional groups that they contain. Within our experimental design, numbers of species and numbers of functional groups jointly affected productivity, having primary effects of similar strength. Productivity was positively related to increasing plant functional group diversity, with legumes playing a strong role as expected. However, there were often significant interactions with location, indicating that effects of functional groups varied at different sites. There were (nonsignificant) differences in the species richness effects at different sites also, but our analysis revealed that these averaged out to provide a relatively simple overall pattern: productivity declined linearly with diversity on a log scale.

There is probably no single best test of the sampling effect. Instead we performed several different tests. All of these tests reject the sampling effect, *sensu strictu*, as a sole explanation of our results. While dominance of particular species plays a role in generating our productivity patterns, dominance was sometimes by species with low rather than high biomass. Complementary and positive interactions also appear to contribute, although it is harder to separate their effects within our design.

COMPARISONS WITH RELATED STUDIES

Biodiversity manipulation experiments are a relatively recent area of ecological research. To date there have only been three other experiments that are closely comparable with the BIODEPTH project, that is, where plant diversity has been directly manipulated in the field by assembling experimental communities. Two of these experiments have been conducted by Dave Tilman and colleagues at Cedar Creek, Minnesota, U.S.A. (see Tilman et al., chapter 3, this volume). In the first experiment, plant species richness alone was manipulated in prairie grassland plots (Tilman, Wedin, and Knops 1996). A second, larger experiment was then performed to investigate both species richness and the effects of five plant functional groups: C3 and C4 grasses, nitrogen-fixing legumes, other herbs, and woody species (Tilman et al. 1997). The third experiment was performed in a Californian serpentine grassland by Hooper and Vitousek (Hooper and Vitousek 1997; Hooper 1998; Hooper and Vitousek 1998), where they manipulated the composition of four plant functional groups: early-season annuals, late-season annuals, perennial bunchgrasses, and nitrogen-fixing legumes.

The overall log-linear relationship found between species richness and aboveground productivity in the BIODEPTH project appears to roughly match the biomass pattern found by the Cedar Creek Biodiversity II experiment (Tilman et al. 1997) and the similar patterns found for plant cover in the Biodiversity I experiment (Tilman, Wedin, and Knops 1996), although by the fourth year of this experiment this relationship appears to have become more linear (Tilman 1999b). Tilman, Lehman, and Thomson (1997) also reported a positive relationship between aboveground biomass and increasing functional group richness that is also in broad agreement with the results reported here. In Hooper and Vitousek's work, each of their five plant functional groups is always represented by two or three species and so they do not therefore examine the effects of species richness. In contrast to the Cedar Creek and BIODEPTH studies, higher levels of plant functional group richness did not increase aboveground productivity. Instead, a single functional group, the perennial bunchgrasses, produced the largest biomass.

Why, given dominance by one functional group (Hooper 1998), did the sampling effect not lead to a positive relationship between productivity and functional group richness? It appears that the species with the largest biomass, the perennial bunchgrasses, were not the best competitors; rather mixtures were dominated by early season annuals, the group with the lowest aboveground biomass. Hooper and Vitousek interpret competition in their communities in terms of Tilman's "R star" (R^*) hypothesis in which the best competitor is the species that reduces limiting resources to the lowest level. Soil nitrogen pool sizes were indeed lowest in mixtures containing early season annuals and equal to levels found in plots with only this group. This would actually lead us to expect an alternative sampling effect in which productivity would decline with increasing competitive dominance by the least productive group. This still leaves the question of why, contrary to the sampling effect model, the group with the lowest aboveground biomass was competitively dominant? It might be that there is a trade-off in allocation to growth versus reproduction, with the annual species allocating heavily to seed (Hooper 1998; Hooper and Vitousek 1998). While early-season annuals might remain dominant in the longer term, perhaps this is a "successional niche" effect (Pacala and Rees 1998) and annuals would eventually be out-competed, as perennial bunchgrasses become dominant. If so, the alternative sampling effect could be a transient effect (see Pacala and Tilman, chapter 7, this volume).

To what degree can productivity patterns in these three experiments be explained by the sampling effect? The sampling effect no doubt plays a role in generating patterns in all of the experiments discussed above (Hooper and Vitousek 1997; Hector et al. 1999; Tilman 1999b) but is it sufficient to explain the results or is there evidence for additional complementary and positive interactions? Tilman and colleagues (Tilman et al. 1997) used regression analysis to study the response of individual species across the diversity gradient and found that many species inhibited themselves more in monoculture and low-diversity plots than they were inhibited by other species in high-diversity mixtures, consistent with the existence of complementary and positive species interactions. In Biodiversity II, by 1998 total biomass of the most diverse plots was two to three times that of the most productive monocultures, suggesting substantial overyielding (see Tilman et al., chapter 3, this volume). Hooper and Vitousek used an analogous procedure, but calculating relative yield totals for each mixture. Although many of their mixtures had RYTs > 1, and overall resource use was positively related to diversity (Hooper and Vitousek 1998), both supporting the occurrence of complementary and positive interactions, the effects did not appear to be strong and may have increased with diversity largely due to the positive effects of nitrogen-fixing legumes in the most diverse mixtures. As discussed above, RYT > 1 does not necessarily mean that mixtures will yield more than the most productive monoculture (Hector 1998), and there was clearly no overyielding in Hooper and Vitousek's experiment. Complementarity and the negative sampling effect would have counteracted each other here, illustrating the point that both the biomass patterns and even derived indices such as RYTs are the net outcome of a combination of positive and negative interactions (Loreau 1998b). We tested the sampling effect in the BIODEPTH project in several ways, including a regression approach similar to Tilman and colleagues and RYTs as used by Hooper and Vitousek. All of our tests rejected the sampling effect as a sole explanation for the observed biomass patterns, consistent with a role for complementary and positive interactions. In summary, all three studies produced evidence for complementary and positive species interactions, at least in some mixtures of plant species. However, in Hooper and Vitousek's work the effects do not appear to have been strong or widespread enough to counteract the effects of dominance of low-yielding annual species that resulted in no overall response of aboveground productivity to increasing functional group richness. Complementary and positive species interactions appear to play a stronger role in generating the results from Cedar Creek and the BIODEPTH project. Better identification and separation of the effects of these different biodiversity processes will enhance our interpretation of these experiments, and these clearly remain as major goals in this area of research at the moment.

Relationships within and between Sites

The positive relationships between biodiversity and ecosystem functioning seen in some of the studies discussed here appear, on the surface, to contradict other known relationships between diversity and productivity. For example, it is well established that fertilizing experimental plots generally increases their productivity and decreases their diversity, and that negative diversity–productivity correlations are also found at larger scales, sometimes as part of a "humpbacked" relationship (Grime 1979; Tilman and Pacala 1993; Huston 1994). Can these apparent contradictions be reconciled? Recent theoretical work by Loreau (1998a) illustrates one possible resolution. Loreau demonstrates, using a mechanistic model of plant competition for a limiting soil resource, how correlations between diversity and productivity seen when comparing between different sites do not have to be the same as the relationship between diversity and productivity seen within sites. For example, with complementarity between species the relationship between diversity and productivity within a site can be positive, even when the relationship between sites is negative. Biodiversity manipulation experiments conducted at multiple locations will be essential for testing what the relationship between the two patterns is in reality.

Individual site analyses, which have less replication and power than the overall analysis, produced a variety of results mostly differing from the overall log-linear pattern. Some of the deviations from the overall log-linear pattern appear to be reduced by the third year of the experiment (Hector et al. 1999) and may be partially explained as a transient effect caused by the relatively low sowing density of particular species in the most diverse plots. Because we are manipulating numbers and types of both species and functional groups, but our level of replication of each level of diversity within a site is relatively low, strong effects of particular species compositions may explain the more complex patterns we see at individual sites. Clearly one of our major tasks now is in determining whether deviations from the overall log-linear pattern at individual sites reflect real ecological differences or variation in the details of our experimental designs.

SUMMARY

The use of standardized experiments replicated at multiple sites, together with comparisons with similarly designed studies, has enabled us to start to identify important generalities and variations in relationships between biodiversity and ecosystem functioning. We are also now beginning to be able to explain why we see this generality and variation in terms of the relative contributions of the processes that generate relationships between biodiversity and ecosystem functioning. This has come about through the development of methodologies for identifying the relative contribution of the influence of dominance of species with particular traits and of the effects of complementary and positive interactions between species.

94

ACKNOWLEDGMENTS

I am very grateful to my colleagues from the BIODEPTH project for allowing me to use their data for this chapter and for their input and comments. I am particularly grateful to Bernhard Schmid for teaching me much about statistical analysis and to John Lawton and Michel Loreau for their help in developing ideas on how to separate different biodiversity effects. Thanks to Paul Giller, Asher Minns, Jasmin Joshi, and Eva Spehn for thorough readings of the text. The BIODEPTH project was funded by the European Commission within the Framework IV Environment and Climate Programme (grant ENV-CT95-0008).

CHAPTER FIVE

Autotrophic-Heterotrophic Interactions and Their Impacts on Biodiversity and Ecosystem Functioning

Shahid Naeem

The living world is, on the protoplasmic level, a
single realm; but it is also the realm of immense
diversity, of many lines of evolution and more
than a million species.
—R. H. Whittaker, 1957

INTRODUCTION

Whittaker (1957) observed that at the most basic level, the biota can be considered a single protoplasmic realm. To transform a geochemical model of ecosystem functioning into a *bio*geochemical model requires adding this "protoplasm," whose metabolic processes modify geochemical processes. A logical starting point for a biogeochemical model is to add a photosynthetic protoplasm, a "green slime," that couples biotic with abiotic carbon cycling. Such a model becomes even more informative when nitrogen, carbon's intimate biogeochemical partner, is included. To treat the "immense diversity" that Whittaker refers to, however, requires additional steps. First, the green slime must be partitioned into trait groups or species clustered by biogeochemical traits. Such trait groups might reflect differences in rates of

96

nutrient uptake, water use efficiency, CO_2 assimilation, associations with N-fixing symbionts, or other functional characters (Chapin et al. 1996). Exploring the consequences of diversifying a green slime model is an important focus of this volume.

Diversified green-slime models, however, are depauperate representations of global protoplasm, and this chapter evaluates how we might improve biogeochemical models by the inclusion of heterotrophic diversity. Based on the fundamental ecological divisions of autotrophy and heterotrophy, Whittaker (1957) recognized that the global protoplasm was at the least made up of three major classes of species defined by three distinct nutritional modes. These modes were photosynthesis (Plantae), ingestion (Animalia), and absorption (Fungi). It is worth noting that the biological basis for Whittaker's divisions—photosynthetic, absorptive, and ingestive—is based on a fundamental distinction between organisms with and without cell walls. Whittaker lumped the unicellular creatures that exhibited all modes of nutrition into one evolutionary basal group, the Protista, a biotic grab bag that he and Lynn Margulis later divided into the Monera and Potoctista (Whittaker 1959; Whittaker and Margulis 1978). Green slime (i.e., plants and algae), however, represents only one-quarter of described species (Hamond 1992), a fraction that is probably inflated due to underrepresentation of invertebrates, fungi, nonphotosynthetic protists, prokaryotes, and viruses (Stork 1997). The other three-quarters of estimated global protoplasmic diversity consists of heterotrophic species. The exclusion of such a dominant fraction of the biota in a green-slime or producer-only biogeochemical model is likely to limit the utility and predictive power of such a model (this volume, chapter 11 by Holt and Loreau; chapter 12 by Balser, Kinzig, and Firestone).

Theoretical and empirical studies support strong roles for heterotrophs and autotroph–heterotroph interactions in ecosystem functioning. First, the bulk of the earth's biota and the bulk of biogeochemical processes are driven by heterotrophic species, not autotrophs. Second, only decomposer heterotrophs, primarily species of Archaea, Bacteria, and Fungi, have the biochemical machinery to transform complex organic materials to inorganic forms (Fenchel, King, and Blackburn 1998). This organic–inorganic material transformation is a critical link in biogeochemical cycling without which dead organic materials would accumulate irrevocably (with the exception of slow, abiotic degradative processes, such as weathering or UV degradation) (Schlesinger 1997). Third, heterotrophs may regulate autotrophic biomass (Hairston, Smith, and Slobodkin 1960; Hairston and Hairston 1993; Oksanen, Power, and Oksanen 1995; Naeem and Li 1998). Fourth, heterotrophs regulate rates of cycling (Loreau 1995; Grover and Loreau 1996). And finally, heterotrophs may regulate system stability (De Angelis 1975; Pimm and Lawton 1977; Pimm 1982; De Angelis 1992; Lawler and Morin 1993; De Ruiter, Neutel, and Moore 1995). The roles of heterotrophs are clearly diverse, important, and well known.

FUNDAMENTALS

Some basic principles are reviewed in this section to clarify terminology and objectives of the chapter. This section further serves to provide the framework used for reviewing current evidence for the importance of heterotrophic diversity in ecosystem functioning.

Classes of Trophically Defined Functional Groups

The earth's biota is represented by four basic classes of functional groups. These four groups are based on the ways in which organisms acquire energy (transformation of light or chemical energy) and the forms of carbon they consume (organic or inorganic). This creates the 2×2 classification scheme that yields the four fundamental groups. Organisms

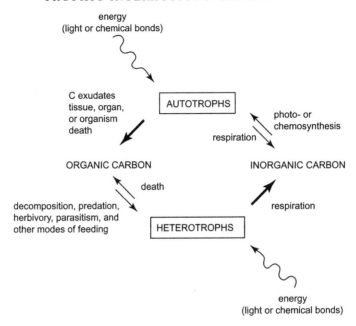

FIGURE 5.1. Fundamental trophic groups.

can be heterotrophs (consumers of organic C) or auto-trophs (consumers of inorganic C), which either acquire energy from light (photoautotrophs and photoheterotrophs) or acquire energy from chemical sources (chemoautotrophs or chemoheterotrophs). Figure 5.1 illustrates these fundamental groupings. Of the four possible combinations, the dominant forms in most ecosystems are photoautotrophs (most plants and algae) and chemoheterotrophs (most protozoa, bacteria, fungi, and metazoa).

The value of such a simple classification scheme is that it reduces what may appear to be an intractable degree of trophic complexity typically seen in communities to a tractable set of major players in ecosystem processes. Here we will consider primarily photoautotrophs and chemoheterotrophs. Further classifications of chemoheterotrophs, however, is necessary for considering additional trophic levels. For simplicity, I will refer to photoautotrophs as *producers*; chemoheterotrophic absorptive, organic–inorganic matter transformers as *decomposers*; and chemoheterotrophic ingestive, organic–inorganic matter transformers as *consumers*. Several ecosystem processes are associated with each of these groups, including decomposition, immobilization, mineralization, production, nutrient uptake, and mortality.

Consumers themselves are conveniently divided into two major groups depending on the basal group from which they acquire their organic carbon sources. The term *basal* stems from the tradition of visualizing pyramids of biomass or pyramids of numbers with consumers at the pinnacle and producers at the base. A more useful representation, however, regards both producers and decomposers as basal groups. For example, a plant fed on by an herbivorous insect fed on by an insectivorous bird fed on by a bird-feeding hawk represents a producer-derived trophic chain of consumers. Decomposer bacteria fed on by bacterivorous nematodes fed on by nematophagous mites fed on by predatory mites represents a decomposer-derived trophic chain. Note that omnivores, or organisms that consume individuals from two or more trophic groups, though important (Diehl 1993), are often ignored to simplify analyses.

The Producer–Decomposer Codependency (PDC)

As basal groups, trophic structure necessarily begins with producers and decomposers, but these two groups are inextricably linked to one another. An important asymmetry, however, exists between these two groups. Organic carbon and organic nutrients are primarily transformed by decomposers while inorganic carbon and inorganic nutrients are transformed by producers, but decomposers can also use (immobilize) inorganic nutrients. This asymmetry leads to an antagonistic mutualism between producers and decomposers that compete for inorganic nutrients but otherwise need each other (Harte and Kinzig 1993). (Producers have been observed to resorb carbohydrate exudates and nutrients (Killinbeck 1996), but this is currently considered to be a minor part of ecosystem functioning.)

FUNDAMENTAL TROPHIC STRUCTURE

From the above, we can derive the fundamental trophic structure of most ecosystems. At the core reside the producers and decomposers that are the principal drivers of material cycling. From this core emanate chains of consumers, one from the decomposers and the other from the producers. Linear chain length from this core tends to be short, seldom containing three or more groups (Pimm and Lawton 1977; Pimm 1982; Lawler and Morin 1993; Sterner, Bajpai, and Adams 1997). Minimal trophic structure therefore consists of producers and decomposers cycling material between inorganic and organic forms with lateral distributions of materials among consumers that contribute to organic and inorganic matter cycling (figure 5.2). In reality, a more reticulate arrangement (Polis and Strong 1996) or system of trophic guilds (Burns 1989) rather than linear structures might better represent nature. Additionally, omnivores are probably more common than appreciated in multitrophic models (Diehl 1993), but the structure portrayed in figure 5.2 can serve as a useful model for minimal trophic structure.

HETEROTROPHIC DIVERSITY AND ECOSYSTEM FUNCTIONING

Given how tightly the producer–decomposer codependency couples the basal groups in ecosystems, it should not be surprising that anything that impinges upon the performance of either group changes ecosystem processes. Consumers feeding on either group will almost certainly change

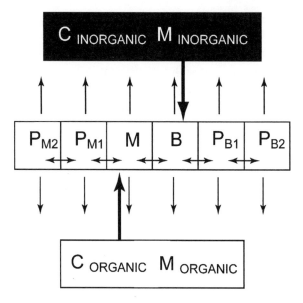

FIGURE 5.2. Fundamental trophic structure. C = carbon; M = microbial decomposers; B = producer biomass; P_{ij} = consumers, where i represents derived from either microbial decomposer biomass (M) or producer biomass (B) in the jth trophic level.

nutrient cycling rates or relative sizes of standing crop, but it is possible that the specific form of the relationship between producer diversity and ecosystem functioning (e.g., an asymptotic relationship) may not be affected by such impacts. A few recent studies, however, suggest that the producer–diversity relationship can be affected by the presence or absence of additional trophic levels. Here, I review several impacts of decomposers and consumers on producers.

Decomposers and Producers Affect Each Other via Carbon Exchange

Observational studies provide abundant evidence for linkages between decomposers and producers mediated by organic carbon exchange. Nonsymbiotic nitrogen fixation, for example, is often enhanced by producers, presumably by providing the large quantities of energy needed by free-living nitrogen-fixing bacteria via organic carbon exudates (Sprent and Sprent 1990). Although difficult to measure precisely, organic carbon exudates of algae (Berman-Frank and Dubinsky 1999) or terrestrial plants (Wall and Moore 1999) support bacteria and fungi (Hobbie 1992). In aquatic systems such organic carbon exudates can have several effects, including protection from UV damage under highlight/low-nutrient conditions, increasing local concentrations of nutrients by sustaining local populations of bacteria, and protecting against viral infections (Berman-Frank and Dubinsky 1999). As mentioned earlier, however, decomposer absorption of inorganic nutrients lends a measure of antagonism to this seeming mutualism.

Harte and Kinzig (1993) have argued that decomposers control this antagonistic mutualism, but only a handful of ecosystems have been analyzed sufficiently to support their conclusions. Studies of interactions between algae and bacteria, for example, support both positive and negative interactions (Jones 1982). Irrespective of the nature of the interaction, empirical evidence supports a clear link between producers and decomposers via the exchange of organic carbon sources.

Consumers Affect the Biomass of Producers and Decomposers

Community ecologists studying heterotrophs concentrate much of their energies on consumers. Beginning with Hairston, Smith, and Slobodkin (1960), understanding the significance of trophic structure began by examining how trophic structure affects the distribution of biomass among trophic levels (Hairston, Smith, and Slobodkin 1960; Hairston and Hairston 1993). The shape of these distributions are seen as multimodal, virtually square-wave distributions, where biomass abundance along a trophic axis occupies two states. Either the basal group (generally photoautotrophs) and every second group above it contain the bulk of the biomass ("green worlds") or the opposite is true ("barren worlds"). Long before Hairston, Smith, and Slobodkin, R. E. Lindeman (1942) pointed community ecologists to the fact that biomass distributions among trophic levels were driven by the dynamics of populations feeding upon one another, with a prominent role for the microbes ("ooze") in ecosystems, though his emphasis was on aquatic systems.

Lindeman may have been aware that his construct was a bit forced, because he recognized inconsistencies in his model (Burns 1989), but the tremendous appeal of discrete trophic groups, for their symmetry, heuristic value, and aesthetics, may have led him and contemporary ecologists to continue to favor strict trophic levels over more reticulate, much more realistic structures, such as Burns's (1989) trophic guilds.

Numerous studies have examined the impact of consumers, nutrients, and other factors on the distribution of biomass among trophic levels (Elliot et al. 1983; Liebold and Wilbur 1992; Abrams 1993; Carpenter and Kitchell 1993; Hairston and Hairston 1993; Carpenter et al. 1996). The continuing debate need not concern us here. The important point is that the number of trophic levels can affect the distribution of biomass among producers, decomposers, and consumers (Naeem and Li 1998).

Trophic Structure Influences Rates of Material Cycling

Food web or trophic models are frequently nonecosystem (e.g., Pimm 1982) in the sense that they do not consider nutrient flows. Instead, they focus on population dynamics of consumers and their prey. Such models, however, can be readily modified to accommodate the principles of nutrient cycling and nutrient dynamics (e.g., DeAngelis 1992). When nutrient cycling is included, the impact of consumers may modify rates of cycling (Loreau 1995), can increase local stability (DeAngelis 1992), and can buffer dynamics against external inputs (Loreau 1994). The inclusion of decomposers can have different effects depending on whether models are donor controlled or Lotka-Volterra (Zheng, Bengtsson, and Ågren 1997). Experimental support of these compelling theories, however, is lacking.

Heterotrophic Diversity Affects Levels and Stability of Ecosystem Processes

Naeem and Li (1997, 1998) used microbial microcosms to explore the relationship between heterotrophic diversity and ecosystem functioning. Each microcosm in this study contained freshwater protist media, a single green algal species (*Chlamydomonas reinhardtii*) as the producer base, and at least three species of freshwater bacteria (*Serratia marcescens, Bacillus subtilis,* and *Bacillus cereus*), which served as decomposers. The only factor that varied was the composition of the consumer community, which consisted entirely of species of protists that fed either on the decomposers, the producers, or both (table 5.1). This study demonstrated that variation in consumer (omnivorous, autotrophic-, or decomposer-derived) species richness showed a strong, negative association with standing autotrophic biomass (figure 5.3). Three principles emerged from this study. First, autotrophic biomass declines in the presence of decomposers, possibly due to competition for nutrients between producers and decomposers. Second, bacterivores increase autotrophic biomass, possibly by suppressing bacteria. Third, consumers that do not feed on decomposers either directly decrease standing autotrophic biomass by herbivory or indirectly decrease standing autotrophic biomass by consuming bacterivores. This is a variation on the Hairston, Smith, Slobodkin theme.

In their second experiment, Naeem and Li (1997) demonstrated that the reliability of heterotrophic suppression of autotrophic biomass was dependent on how many heterotrophic species were found per functional group. The same

105

TABLE 5.1. Species Composition of Microbial Microcosms Used in Experiments

Species Richness	Bacterivores			Omnivores			Carnivores		
	CHIL	COLP	SPIR	PA	PC	PM	EUPL	APROT	CHAO
1	X								
1		X							
1							X		
1								X	
1									X
2	X			X					
2	X					X			
2		X				X			
2	X	X							
3	X			X	X				
3	X	X					X		
3	X			X					X
3	X					X			X
3	X			X				X	
3	X					X		X	
4	X	X		X	X				
4	X	X		X			X		
4	X	X				X	X		
4	X			X	X				X
5	X	X		X	X				X
5	X	X		X			X		X
5	X	X		X		X	X		
7	X	X	X	X		X	X	X	X

Notes: All microcosms contained *Chlamydomonas reinhardtii* (photoautotroph, unicellular alga), *Serratia marcescens, Bacillus subtilis,* and *Bacillus cereus* (decomposers, bacteria). Species are abbreviated as follows: Detritivores: CHIL = *Chilomonas* sp., COLP = *Colpidium* sp., SPIR = *Spirostomum* sp. Consumers of either autotrophs or detritivores or both (omnivores): PA = *Paramecium aurelia,* PC = *P. caudatum,* PM = *P. multimicronucleatum.* Predators of consumers: EUPL = *Euplotes* sp., APROT = *Amoeba proteus,* CHAO = *Chaos carolinensis.* Containers were sterile, disposable, polystyrene petri-dishes. Media consisted of sterile Carolina Biological Supply protozoan media and three sterilized wheat seeds.

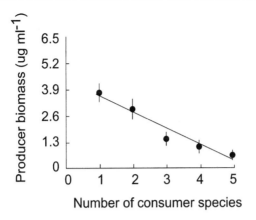

FIGURE 5.3. The relationship between consumer heterotrophic species richness and standing producer biomass. Circles represent means and error bars represent one SE of unicellular algal biomass measured in microbial microcosms. The number of heterotrophic species refers to the number of either decomposer-derived or producer-derived heterotrophic species of protists (flagellates or ciliates). This figure illustrates that increases in consumer protistan diversity, here measured as species richness, can reduce standing algal biomass in a predictable manner to very low levels. See Naeem and Li (1998) for details. After (Naeem and Li 1998).

microcosm design was employed, but the number of functional groups and number of species per functional group was varied systematically. The number of trophically defined functional groups varied from a minimum of two (bacteria and algae) to a maximum of five (bacteria, algae, bacterivorous protists, omnivorous protists, and carnivorous protists). The number of species per functional group (except for carnivorous protists, where maintaining a third species within the microcosm was not possible) varied from one to three. Local extinction occurred in most microcosms, ensuring that replacement by substitutable species within functional groups could occur. The experiment was replicated at two different light levels and two to three different nutrient levels to provide a more robust test of the patterns observed.

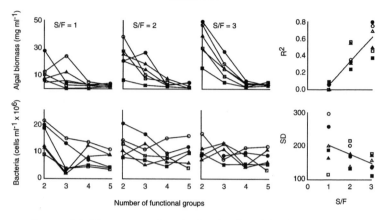

FIGURE 5.4. The pattern of production is affected by the number of species per trophically defined functional group of protistan consumer species. Top, left: Algal production (biovolume) at the end of the experiment as a function of functional group richness. Top, right: The R^2 of linear regressions of $\log_{10}$ (algal production) against numbers of protistan species per trophically defined functional group in microbial microcosms. This figure shows that predictability of algal production increases sharply as numbers of species (S) per functional group (F) increases. Note decline in production is a function of functional group richness, which is similar to results obtained in earlier experiments (figure 5.3). Bottom, left: Bacterial biomass at the end of the experiment as a function of number of functional groups in microcosm. Bottom, right: The standard deviation of microbial production plotted against species per functional group. This figure shows that bacterial production varies less, or is more predictable, when S/F increases. Open symbols are high-light replicates while closed symbols are low-light. Circles are full-strength media, triangles are half-strength, and squares are one-quarter nutrients. Each symbol represents the mean of four replicates. After Naeem and Li (1997).

The same decline in producer biomass with increasing trophic diversity was observed in this study as in the earlier experiment (figure 5.3), but the more species per functional group, the more predictable the response (figure 5.4). This experiment provided general support for the principle of ecosystem reliability (Naeem 1998; Rastetter et al. 1999). A similar multitrophic-level study (McGrady-Steed, Harris, and Morin 1997), which did not directly manipulate trophic structure but examined ecosystem functioning more closely, found that both resistance to invasion and ecosystem predictability increased with increasing diversity.

Like decomposers, consumer species richness may also influence consumer biomass, but this possibility is relatively unexplored. To my knowledge, the only explicit manipulation of consumer diversity is Norberg's (2000) study, which directly manipulated cladoceran, zooplankton species richness and demonstrated an idiosyncratic response of zooplankton biomass to species richness. This study was a microcosm study and had fewer than four species (Norberg 2000), but it does suggest that consumer diversity may affect consumer biomass.

Heterotrophs Modulate Producer Diversity Effects

Recent studies of producer diversity have shown that diversity and production are positively associated with a number of ecosystem processes (Naeem et al. 1994; Naeem et al. 1995; Naeem et al. 1996; Tilman et al. 1996; Hooper and Vitousek 1997; Tilman et al. 1997; Tilman, Lehman, and Thomson 1997; Hooper 1998; Hooper and Vitousek 1998). Such studies argue for a role of biodiversity in biogeochemical processes even though only one study (Naeem et al. 1994) manipulated the diversity of more than one trophic level. Recent evidence is accumulating, however, that heterotrophs interact with producers and modify the impacts of variation of producer diversity on ecosystem processes. For example, both mycorrhizal fungal diversity (Van der Heijden et al. 1998) (figure 5.5A) and insects (Mulder et al. 1999) modify producer diversity effects on ecosystem production (figure 5.5D). CO_2 flux, a product of producer, decomposer, and consumer metabolic activities, showed a positive relationship to microbial species diversity in freshwater microcosms (figure 5.5B). Finally, variation in producer diversity that did not yield aboveground responses did exhibit microbial responses, such as a positive relationship between

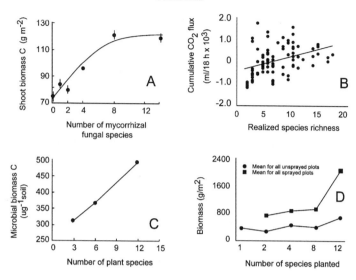

FIGURE 5.5. Examples of how heterotrophs interact with producer diversity. After (A) McGrady-Steed, Harris, and Morin 1997; (B) Chapin et al. 1998; (C) Van der Heijden et al. 1998; (D) Mulder et al. 1999.

microbial biomass (most likely entirely due to decomposer microbes) and plant species richness (Chapin et al. 1998) (figure 5.5C).

The most complex effects are shown in a recent freshwater microbial microcosm study (Naeem, Hahn, and Schuurman 2000), where varying decomposer diversity (bacterial species richness) and varying producer diversity (algal species richness) were both shown to affect producer and decomposer biomass production (figure 5.6). That is, both the main effects (producer and decomposer diversity) and the interaction between the two were significant using a two-way ANOVA ($P < 0.05$) (Naeem, Hahn, and Schuurman 2000). More importantly, variation in decomposer diversity nearly doubled (1.82 times) the range of producer biomass observed in microcosms where only producer diversity was varied. These studies suggest that variation in producer diversity by itself is insufficient to account for variation in pro-

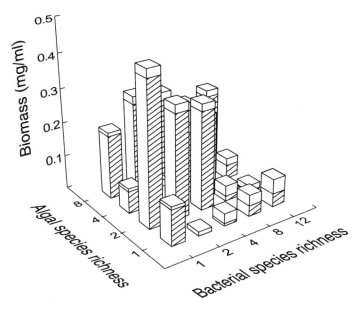

FIGURE 5.6. The relationship between biomass production and simultaneous variation in producer and decomposer diversity. Boxed top portions of columns represent bacterial (decomposer) biomass while lower portions represent algal (producer) biomass in 50 ml freshwater microbial microcosms after four weeks. After Naeem, Hahn, and Schuurman 2000.

duction due to biodiversity loss. Decomposer and consumer diversity, and the *interaction* between these and producer diversity, may generate differences in ecosystem processes.

Summary of Empirical Findings

From the above, the best-documented impacts of heterotrophs on producer-driven biogeochemical processes concern the interactions between decomposers and producers and the impacts of trophic chains on these basal trophic groups. The theoretical and empirical studies surveyed above, though an eclectic and scant assemblage of studies, point to a variety of possible effects that heterotrophs may

111

have on producer-driven processes. These effects provide motivation for modifying the basic producer-only model that is the focus of this book. They also provide some hints of what sorts of effects one might expect. These include much greater variation in model predictions of producer biomass and ecosystem processes tied to autotrophic production and decomposition rates due to a variety of simple effects (e.g., competition between decomposers and producers for inorganic nutrients) to complex effects (e.g., interactions between variation in producer and decomposer diversity). The next section offers some suggestions for multitrophic-level model development.

IMPLICATIONS FOR AUTOTROPH-ONLY MODELS

Useful autotroph–heterotroph model development requires extracting the fundamental, common set of features typical of multitrophic systems from the mountain of detail concerning the various ways heterotrophic species affect producers. From the brief review of trophic groups above, the simplest yet most realistic model would contain the following universal elements:

1. photoautotrophs (producers)
2. decomposers
3. a two-link chain of decomposer-derived consumers
4. a two-link chain of producer-derived consumers

The basic model used in this volume (presented in chapter 8) simplifies the treatment of material by tracking only C, N, and H_2O (W) as these materials are cycled among the organic and inorganic pools. C is compartmentalized into refractory (fast) and recalcitrant (slow) fractions. The inclusion of heterotrophic effects would minimally require the inclusion of decomposers and consumers. This would partition the living biomass into several compartments, each of which needs to account for N and C cycling and water effects on growth.

Decomposers

The simplest approach to incorporating decomposers is to redefine decomposition rates in terms of microbial processes and minimally treat decomposer diversity as two "species" of decomposers. In the common ecosystem model, this is accomplished by assuming that each plant produces at least two classes (fast and slow) of organic C, the C:N ratio being the critical determinant of rate of decomposition of either class. Decomposers could be minimally divided into two groups, one group would be those that specialize on recalcitrant organic material but at a cost to growth, while the second group makes use of the labile fraction with higher growth rates, which is done in chapter 12 (Balser, Kinzig, and Firestone). From the principle of producer–decomposer codependency, we assume microbes are primarily carbon limited.

Trophic Levels

Two trophic levels should be added to each basal group. Herbivores (P_{B1}) and carnivores (P_{B2}) feed on plants while bacterivores (P_{M1}) and predators of bacterivores (P_{M2}) are added (figure 5.2). To consider the impact of trophic groups on microbes we can assume that both forms of microbial decomposers are edible and have the same C:N ratio. This is accomplished by different rates of metabolic loss of C, where M_s respires more C. Herbivores are considered in chapter 11 (Holt and Loreau).

Material Pools

The inclusion of two specialist classes of microbial decomposers (above) requires a specific treatment of N as two classes of compounds. The common ecosystem model approach used in this volume divides N into two forms: N_f

associated with organic matter that decomposes quickly and N_s associated with organic matter that decomposes slowly. Ultimately both forms accumulate in the inorganic N pool by mineralization. One problematic twist is that microbes can convert inorganic N into microbial biomass (immobilization), a factor that is regulated by C availability.

Nitrogen is complicated. In addition to terms for deposition and leaching in the basic model, we need to add terms that reflect the contributions by mineralization from consumers and decomposers. The consumer terms are readily added by assuming that respiration is related to mineralization. Additional modifications involve the inclusion of decomposer immobilization of inorganic N and the replacement of rates of decomposition by decomposer conversion, as with C. Thus, N would be a function of gain due to deposition, loss due to leaching, loss due to plant uptake, gain due to heterotrophic mineralization, and loss due to immobilization. Loss due to denitrification is minimal, but for wet, anaerobic soils, tracking loss by this route would be important. Balser, Kinzig, and Firestone (chapter 12) employ this approach in their analyses.

DISCUSSION

Two common perspectives among modelers is to either consider heterotrophic diversity as fat on the autotrophic backbone of ecosystems or, in sharp contrast, consider autotrophs as merely fodder for ecosystem processes conducted primarily by heterotrophs. Neither perspective works well because autotrophs and heterotrophs are mutually codependent (Harte and Kinzig 1993; Naeem, Hahn, and Schuurman 2000). Levels, dynamics, and the stability properties of ecosystem functioning are governed by both autotrophic diversity and heterotrophic diversity, as well as the complex interactions between the two.

Trophic differences among species represent the basis for

evolutionary divergence among the plants, fungi, animals, and microbes, so it should be no surprise that ecosystems are strongly affected in different ways by such different species. These groups, defined by different modes of nutrition, are biologically distinct (e.g., possess or lack cell walls, can or cannot produce specific exo- or endoenzymes) and are not substitutable.

From this survey of empirical studies, several principles emerge. First, decomposers are locked in an "antagonistic mutualistic" relationship with producers (Harte and Kinzig 1993), which leads to inseparable responses of these groups to environmental variation. Second, consumers affect rates of movement of materials among different pools (Loreau 1995; Grover and Loreau 1996). Third, consumers determine the distribution of biomass among trophic groups (Hairston, Smith, and Slobodkin 1960; Del Giorgio and Gasol 1995; Naeem and Li 1998). Fourth, stability and reliability, both of systems and populations, are affected by trophic structure and numbers of species within trophic groups (Pimm and Lawton 1977; Lawler and Morin 1993; McGrady-Steed, Harris, and Morin 1997; Sterner, Bajpai, and Adams 1997). Finally, interactions may exist between diversity at one level and diversity at another (Fig. 5.6) (Harte and Kinzig 1993; Naeem, Hahn, and Schuurman 2000).

Theoretical and empirical evidence suggest some possible outcomes of such an autotroph–heterotroph model. First, the addition of trophic groups most likely enhances overall rates of cycling by increasing flows of biomass to the dead organic pool and increasing rates of mineralization. These enhanced rates will be accompanied by lower standing crops of autotrophic and decomposer biomass, but higher rates of biogeochemical processes. Whether or not this enhancement of rates will translate into differences in the biodiversity–ecosystem functioning relationship remains to be seen.

Second, given the biogeochemical linkages among trophic groups, the impacts of varying producer diversity on ecosystem functions, such as enhanced production and nutrient drawdown (Naeem et al. 1994; Naeem et al. 1995; Naeem et al. 1996; Tilman, Wedin, and Knops 1996; Hooper and Vitousek 1997; Tilman et al. 1997; Tilman, Lehman, and Thomson 1997; Hooper 1998; Hooper and Vitousek 1998), are likely to have feedbacks through other trophic levels. The mechanistic basis for these feedbacks is likely to be the exchange of organic carbon between producers and decomposers.

Third, variation in diversity in one trophic level is likely to modulate the impacts of variation in another trophic level. For example, if increasing producer diversity increases the efficiency of light, water, and nutrient exploitation in a limited space, then increasing decomposer diversity may lead to similar increases in trophic exploitation of resources. Increases in decomposer diversity and producer diversity would intensify the competition for and reduce the standing level of inorganic nutrients.

It is important to note that there is little question that decomposers and consumers are important in regulating ecosystem functioning, but the importance of biodiversity within these groups, the focus of this chapter, is much more difficult to assess. One possibility is that decomposer communities are less sensitive to variation in their biodiversity compared to variation in producer and consumer biodiversity. Exoenzymatic capabilities are widespread among microbial taxa in decomposer communities, and mutation and gene exchange, coupled with high reproductive rates and large population sizes, suggests that biodiversity loss may not correlate strongly with loss of functional diversity. Functional capabilities may be widespread among taxa or may return quickly even if lost due to habitat destruction. It is possible that recovery is much faster in decomposer than producer and consumer communities because microbes typically exhibit order-of-magnitude higher rates of ecological and evolutionary processes. Freshwater and marine systems driven by microbial producers as well as microbial decomposers may set these ecosystems apart from metaphytan-driven terrestrial ecosystems. For this reason, focusing on consumers may prove more valuable for predicting ecosystem response to biodiversity loss.

The inclusion of trophic groups in biodiversity–ecosystem function studies has begun, but this area of research clearly needs expansion. Both current experimental research and theory (Balser, Kinzig, and Firestone, chapter 12; Holt and Loreau, chapter 11) support potential roles for heterotrophs in modifying biodiversity–functioning relationships, but the role of heterotrophic diversity itself is still unclear. What are the roles of decomposer diversity? What are the roles of autotrophically derived consumers? What are the roles of decomposer-derived consumers? What is the role of trophic structure either in decomposer or autotrophic-derived food chains? Finally, what are the interactions among these factors? The promise of insights into these and other issues strongly encourages multitrophic approaches to understanding the ecological consequences of changing patterns in biodiversity.

Biodiversity experiments that manipulate plant diversity are difficult, and simultaneously manipulating heterotrophs and producers makes this already difficult line of ecological research even more difficult. Perhaps the most straightforward approach is a "brute-force" simultaneous manipulation of plant, insect, and microbial diversity in a well-replicated series of experimental plots. Such an experiment would require several hundred replicates for examining functional group richness and perhaps thousands of replicates for examining species richness within both autotrophic and heterotrophic functional groups. Though inelegant, it would provide answers to the questions raised above.

There are, however, more focused and tractable experiments one could conduct. For example, field experiments that manipulate functional group richness of heterotrophic diversity on simple autotrophic communities, such as guilds of specialist insect herbivores on single, clonal host species, may prove valuable as a starting point. One could test whether autotrophic production is affected by the number of functional groups (e.g., sucking insects, chewing insects, predatory, parasitoid, or other trophically defined groups) or the number of species per functional group. The sampling effect, heterotrophic niche complementarity, and tests of stability and reliability of mineralization, community respiration, or biomass production (heterotrophic or autotrophic) could be measured as response variables in such experiments.

Decomposer communities are more difficult to manipulate under field conditions, but soil could either be partially sterilized by varying amounts of chloroform fumigation, selective bactericides and fungicides, or microwave treatment. Alternatively, completely sterilized soil could be reinoculated with live soil plugs from different habitats (high-diversity treatment) or single habitats (low-diversity). Quantification of microbial diversity could be accomplished by a variety of methods (Zelles et al. 1995; Borneman and Triplett 1998; Torsvik et al. 1998) to ensure that treatments were effective. Bacterivorous and fungivorous microarthropod communities may be similarly manipulated by use of fauna-infected litter or soil plugs placed in sterilized soil.

Uncovering interactions between diversity at one trophic level and diversity at another level can follow the design we employed in our microbial microcosm research (Naeem, Hahn, and Schuurman 2000), but this is not an elegant solution. A more effective solution might be to partition trophic manipulations to better explore the linkages. For example, establishing plant communities at different diversity levels in sterilized soil that had been previously maintained at different levels of decomposer diversity would provide a way to isolate decomposer diversity effects from autotrophic diversity effects. Such "conditioned" soil would exhibit the effects of carbon-depleted decomposer communities on nutrient pools. Theoretically, the most diverse communities would have exhausted the most diverse set of carbon sources, thereby leading to larger inorganic nutrient pools than might be found in soils that had depauperate decomposer communities. Plants would respond favorably to soil that had been exposed to high-diversity decomposer communities, and higher-diversity plant communities may respond even more favorably by exhibiting higher production.

Conceptually, constructing effective experimental designs is not difficult, but logistically, due to the combinatoric explosion of necessary replicates to explore variation in even modest amounts of diversity, multitrophic biodiversity–functioning experiments are likely to require larger scale efforts than ecology has traditionally employed in field experiments. Microcosm and mesocosm experiments are ready solutions, but field experiments are necessary as robust tests of microcosm findings. Elegant solutions to the logistic problems of multitrophic diversity–functioning experiments may emerge from theoretical explorations such as those by Balser, Kinzig, and Firestone (chapter 12) and Holt and Loreau (chapter 11). Perhaps, as this volume suggests, theory may provide focus and testable hypotheses for empirical study.

CHAPTER SIX

Empirical Evidence for Biodiversity–Ecosystem Functioning Relationships

Bernhard Schmid, Jasmin Joshi, and Felix Schläpfer

INTRODUCTION

The implementation of ecosystem models relating biodiversity to ecosystem functioning requires empirically derived parameter values. We assemble and summarize these parameter values by reviewing empirical studies. From this review, it becomes obvious which parameters have good empirical estimates and where the data are still poor or lacking. Generally, there are good data for aboveground net primary productivity, poorer data for mineralization and decomposition, and hardly any data for water relations. The essential point of the review (as well as of the models) is that we want to know how changing the diversity of species in ecosystems alters the parameters of ecosystem functioning. How changes in diversity are induced by various human activities or global change in the first place, and how these changes relate to experimentally simulated species loss, has not been the primary focus of the reviewed studies and will not be in this review. We will discuss, however, how this important question can be addressed in future studies.

Because the plant community is a major driver of ecosystem models, and because a large number of empirical studies looked at effects of plant diversity on ecosystem functioning, we focus our review on these studies. Further, because the different diversity levels are usually expressed in number of species, assuming equal initial abundances, we classify and synthesize the studies according to the species numbers used in the diversity treatments. We only review a very limited number of observational studies because usually they did not measure the influence of concomitant "third variables," which may affect the diversity–productivity relationship (e.g., edaphic or climatic differences between units of observation; for recent reviews, see Grace 1999; Waide et al. 1999; and Tilman and Lehman, chapter 2, this volume).

Biodiversity effects on ecosystem functioning may be smaller under uniform conditions, and biodiversity may affect average parameter values less than their variances. However, empirical studies rarely included different heterogeneity levels as treatments or variances of ecosystem response variables as measurements. Therefore, we can only briefly review evidence related to these questions. Studies that only report variances of population response variables (e.g., Frank and McNaughton 1991; Sankaran and McNaughton 1999) are not included in our review because variances of ecosystem response variables cannot be derived from them.

How a reduction in the number of species in a community affects a particular ecosystem variable is a conceptually simple question, but this question is difficult to study in the real world. First, there are conflicts in combining random sampling with the need to make experimental communities with realistic properties—for example, communities that are constant in species number and abundances over space and time. All possible communities may not be able to exist in reality. Further, true replication of diversity levels must be achieved to avoid confounding diversity effects per se (number effects) with effects of individual species and individual species-combinations (identity effects). In addition, the degree to which the effects of species richness and plant functional-group richness can be disentangled depends on the experimental design used (see also Allison 1999).

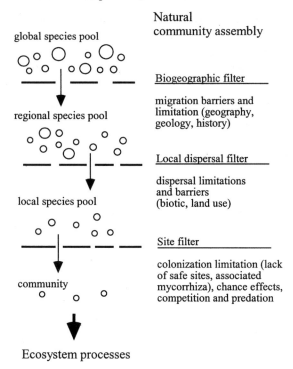

FIGURE 6.1. Processes involved in the assembly of regional and local species pools and local plant communities.

A number of processes at various levels determine the maximum number of species available for community assembly. For example, the maximum number of plant species on a plot of land is dependent on the global species pool from which the regional and then the local species pool is assembled (figure 6.1). However, the actual number of plant species on a plot of land may be lower because of inappropriate site conditions for some species (including the presence or absence of other species). Moreover, the "effective diversity" (Tilman and Lehman, chapter 2, this volume) further depends on the relative abundances of these species, which may differ, for example, due to site-specific competitive interactions and other factors.

In the final section of this review, we pay particular attention to differences among studies in "species filters" or sampling processes invoked for community assembly. We believe that attention to these differences can partly resolve controversies in the literature about the relationship between biodiversity and ecosystem functioning (Aarssen 1997; Huston 1997; Tilman 1997a; Grime 1998; van der Heijden et al. 1999; Wardle 1999). Experimental studies usually use arbitrary combinations of species simulating different-sized local species pools as their diversity treatment, whereas observational studies mostly use the actual species number as their diversity variable. To interpret the results of experimental studies it is therefore important how the local species pool was varied. For the observational studies it is important to know how the site conditions differed among plots, which, as mentioned above, is rarely possible. The combination of local-species-pool manipulations and control of actual species number would be possible in two-stage experiments as suggested later on in this chapter (see also Holt and Loreau, chapter 11, this volume).

PLANT DIVERSITY EFFECTS ON ECOSYSTEM FUNCTIONING

Our selection of papers for this review (see tables 6.1–6.3) was restricted to studies comparing ecosystem-level or community-level variables among plots differing in diversity within the primary producers' trophic level (see also Schläpfer and Schmid 1999). Diversity levels were manipulated directly, and indirectly by varying site environmental conditions (nutrient level, successional stage). Observational studies reporting productivity of biomass were only included if plots were selected according to diversity levels (as the explanatory variable) and the study accounted for among-plot variation in "third variables" (climatic and edaphic con-

TABLE 6.1A. Effects of Plant Diversity on Net Primary Productivity under Equilibrium Conditions

Reference	Diversity Gradient*	Species Comp.†	Climatic Zone‡	Ecosystem Type	Time Scale
Naeem & Li 1997	exp., M	r	—	aquatic	57 d
Naeem et al. 1994/95	exp., P	n	temp	ruderal	206 d
Symstad et al. 1998	exp., P	n	temp	grassland	4 m
	exp., P	rr	temp	grassland	4 m
Naeem et al. 1996	exp., Pt	r	temp	ruderal	65 d
Garnier et al. 1997	exp., Pt	rr	temp	grassland	60 d
Smith & Allcock 1985	exp., F	r/nr	temp	high N gr.	5 y
	exp., F	r/nr	temp	low N gr.	5 y
Tilman, Wedin & Knops 1996	exp., F	r	temp	grassland	2 y
Hooper & Vitousek 1997	exp., F	r	med	grassland	2 y
Tilman et al. 1997	exp., F	rr	temp	grassland	3 y
Hector et al. 1999	exp., F	rr	med, temp	grassland	2 y
Spehn et al. 2000b	exp., F	rr	temp	grassland	3 y
Stocker et al. 1999	exp. F	n	temp	grassland	3 y
Niklaus et al. 2001b	exp., F	n	temp	grassland	5 y
Tilman, Wedin & Knops 1996/Tilman 1990a	obs., F	nr	temp	grassland	—
Troumbis & Memtsas 2000	obs., F	nr	med	shrubland	—

Notes: Time scale refers to the duration of the biodiversity experiment. See the text for the definition of the subject area of this and Tables 6.1B and 6.1C compilations (see also Schläpfer and Schmid 1999).
* exp. = experimentally created diversity gradient; obs. = observed natural gradients; nutr. = gradients created by different nutrient levels; succ. = gradients created by different successional stages; P = phytotron; F = field study; Pt = pot experiment, glasshouse; M = microcosm study.
† r = random mixture; rr = random mixture with restrictions; n = nested composition; nr = other nonrandom mixture.
‡ temp = temperate; med = Mediterranean; trop = tropical.
§ Number of species in observed plots.
‖ Parentheses designate functional group numbers not explicitly stated in study.

ditions). An exception to this rule was made in the compilation of studies reporting biodiversity effects in response to perturbations, where we included additional observational studies that did not follow such a rigorous design.

General Patterns under Uniform Conditions

REVIEW OF EMPIRICAL STUDIES

Most of the 41 empirical studies that deliberately analyzed the relationship between plant diversity and ecosystem functioning were made in temperate grasslands, excluded perturbation treatments, and measured ecosystem performance

TABLE 6.1A. *Continued*

Plot Size (m²)	Diversity Levels§	No. Funct. Groups‖	Min./Max. Average Prod.# (g/m²/y)	Treatment with Min./Max.**	Type of Diversity Effect††
50 ml	1,3‡‡	—	~10/30 mg/ml	1/3	↑
1	2,5,16	(1–2)	—	2/16	↑
0.065	1,4,8,12	1–4	~246/731	8/12	idiosyncratic
0.065	1–10	(1–4)	~492/792	1/10	↑ log-linear
0.03	1–16	(2)	300/433	1/16	↑ log-linear
0.017	1, 4	1–2	dep. on nutr.	1≈4	none
2.5	1,2,7	(1–2)	1050/1270≈1290	1/2≈7	↑ asymptotic
2.5	1,2,7	(1–2)	580/1010	1/2	↑↓ optimum
9	1–24	(1–4)	~33/57%ᵃ	1/24	↑ asymptotic
2.25	2–9	1–4	~180 all	3/1≈2≈4	none
169	1–32	1–5	~75/180	1/16	↑ log-linear
4	1–32	1–3	256/1258	18/16	↑ log-linear
4	1–32	1–3	181/491ᵇ	1/8	↑ log-linear
1.27	5,12,31	3	ᶜ	5/31	↑(2.yr)/none (3.yr)
1.27	5,12,31	3	—	5/31	↑
—	4–23	—	~26/86%ᵃ	6/4	↑ linear
—	1–16	—	~100/525ᵇ	1/9	↑ log-linear or linear

\# Minimum and maximum value of the observed variable (diversity treatment averages); ᵃ total plant cover at max. standing crop; ᵇ at maximum standing crop; ᶜ measured as net ecosystem CO_2 flux.
** Diversity treatments yielding minimum and maximum average values of the observed variable.
†† Presence/absence, direction, and (if > 2 treatment levels) form of the relationship between biodiversity and the observed ecosystem process: ↑ = relationship positive; ↓ = relationship negative, none = no significant relationship; id. = idiosyncratic = significant differences among some diversity treatments, but no simple general relationship *or* effects only at some measurement points in time.
‡‡ Aquatic systems in which only autotroph diversity was varied.

after several months to years as aboveground net primary productivity (tables 6.1, 6.3, 6.4). This measure ranges from less than 100 g to more than 1000 g dry mass per square meter and year, and is generally lowest for an average monoculture and in nearly 50% of the cases highest for the mixed communities that contain 10 or more plant species. Effects are highly significant, and the shape of the relationship is curvilinear with increasing productivity losses per species lost as diversity is reduced, especially in the experiments with comparably large plot sizes and large numbers of replicates at each diversity level (Tilman, Wedin, and Knops 1996; Tilman et al. 1997; Hector et al. 1999). Without

TABLE 6.1B. Effects of Plant Diversity on Processes Related to Mineralization and Decomposition under Equilibrium Conditions

Reference	Diversity Gradient*	Species Comp.†	Climatic Zone‡	Ecosystem Type	Time Scale
Bardgett & Shine 1999	exp., M	n	temp	grassland	100d
Wardle & Nicholson 1996	exp., P	r	temp	gr./ruderal	78–502 d
	exp., P	r	temp	gr./ruderal	78–502 d
Symstad et al. 1998	exp. P	rr	temp	grassland	4 m
	exp. P	n	temp	grassland	4 m
Tilman, Wedin & Knops 1996	exp., F	r	temp	grassland	2 y
Tilman et al. 1997	exp., F	r	temp	grassland	3 y
Wardle et al. 1997	exp., F	r	temp	foliar litter	20–300 d
Hooper & Vitousek 1997/98	exp., F	r	med	grassland	2 y
	exp., F	r	med	grassland	2 y
	exp., F	r	med	grassland	2 y
Scherer-Lorenzen 1999	exp., F	rr	temp	grassland	2 y
	exp., F	rr	temp	grassland	2 y
	exp., F	rr	temp	grassland	1 y
Hector et al. 2000	exp., F	rr	temp	grassland	18 m–21 m
	exp., F	rr	temp	grassland	18 m–21 m
	exp., F	n	temp	grassland	18 m–21 m
Spehn et al. 2000a	exp., F	rr	temp	grassland	3 y
	exp., F	rr	temp	grassland	2–4 y
Niklaus et al. 2001a	exp., F	n	temp	grassland	5 y
	exp., F	n	temp	grassland	5 y
	exp., F	n	temp	grassland	5 y
	exp., F	n	temp	grassland	5 y
Ewel, Mazzarino & Berish 1991	succ., F	nr	trop	plantation, forest succ.	6 y
	succ., F	nr	trop	plantation, forest succ.	6 y

Notes: See Table 6.1A for key; several less recent studies with plant litter are not included in this table; these report mostly idiosyncratic or no effects of species diversity (see Wardle et al. 1997).
§§ Plot averages (Table 1 in Brown & Ewel 1987).

TABLE 6.1B. *Continued*

Plot Size (m²)	Diversity Levels§	No. Funct. Groups‖	Observed Variable	Type of Diversity Effect††
—	1–6	(1–3)	decomposition of community-specific litter in a standardized environment	↑
0.01	1, 2	(1–2)	soil microbial biomass	↑ / idiosyncratic
0.01	1, 2	(1–2)	decomposition rate	idiosyncratic
0.065	1–10	(1–4)	nitrate leaching	none
0.065	1,4,8,12	1–4	nitrate leaching	none
9	1–24	(1–4)	mineral nitrogen in rooting zone	↓ log-linear
9	1–32	1–5	mineral nitrogen in rooting zone	↓ log-linear
0.01	1–8 (pool:32)		decomposition rate; nitrogen release; microbial biomass	idiosyncratic
2.25	2–9	1–4	soil N pools at single time point	↓ log-linear
2.25	2–9	1–4	N content of microbial biomass	↑ / idiosyncratic
2.25	2–9	1–4	N retention over growing season; nitrification; microbial biomass P	none
4	1–16	1–3	net N-mineralization	none
4	1–16	1–3	net N-mineralization of communities containing legumes	↓ log-linear
4	1–16	1–3	decomposition of community-specific litter	↑ linear
4	1–11	1–3	decomposition of standardized litter	↑
4	1–11	1–3	decomposition of community-specific litter in a standardized environment	idiosyncratic
4	1–11	1–3	decomposition of community-specific litter in a standardized environment	↑
4	1–32	1–3	microbial biomass	↑ log-linear
4	1–32	1–3	decomposition of standardized litter	none
1.3	5,12,31	3	soil nitrate pool	↓
1.3	5,12,31	3	nitrogen mineralization	none
1.3	5,12,31	3	nitrification	↑
1.3	5,12,31	3	microbial biomass	idiosyncratic
256	1,40,50,60§§	1, several	soil nitrate pool; cation loss	↓
256	1,40,50,60§§	1, several	ammonium, P, and S pools; N mineralization; organic matter variability	none

TABLE 6.1C. Effects of Plant Diversity on Water Relations
under Equilibrium Conditions

Reference	Diversity Gradient*	Species Comp.†	Climatic Zone‡	Ecosystem Type	Time Scale	Plot Size (m^2)
Hooper & Vitousek 1998	exp., F	r	med	grassland	2 y	2.25
Spehn et al. 2000b	exp., F	rr	temp	grassland	2 y	4
Stocker et al. 1999	exp., F	n	temp	grassland	1 y	1.27

Notes: See Table 6.1A for key.

weighting the studies shown in table 6.1A, the table-wide average productivity loss for each halving of species number is roughly 50–100 $g/m^2/y$.

Experimental studies varying plant diversity or plant litter diversity and reporting mineralization or decomposition rates under nonperturbed conditions do not reveal a coherent pattern of effects (table 6.1B). There is a weak indication that more nitrogen is mineralized and taken up by more diverse plant communities than less diverse ones and that species loss is most significant if entire functional groups such as nitrogen-fixing plants are removed. In the few cases where soil microbial biomass, activity, or decomposition rates were assessed, the effects of increasing plant diversity on these variables were nonsignificant, idiosyncratic, or positive. There are no empirical studies investigating carbon storage in the belowground slow carbon pool in ecosystems differing in plant diversity.

To our knowledge, the influence of diversity on the water relations of plant communities has been explicitly studied in only a few experiments (table 6.1C). One study indicates increased transpiration with increasing plant diversity as soil moisture levels are reduced or leaf-area index is increased (and canopy temperature decreased; E. Spehn et al., un-

TABLE 6.1C. *Continued*

Diversity Levels§	No. Funct. Groups‖	Observed Variable	Min./Max.#	Min./Max. Species**	Type of Diversity Effect††
2–9	1–4	soil moisture at some time points	0.14/0.17 g/g (at depth 10–20 cm)	3/1 fg	↓ / id.
1–32	1–3	LAI, as a surrogate for transpiration	1.13/8.68 m²m⁻²	1/32	↑ log-linear
5,12,31	3	evapotranspiration (H_2O flux)	dep. on light	5 = 12 = 31	none

publ. data) with the number of plant species or functional groups in the community, whereas no effect, an idiosyncratic effect, or even a negative relationship was observed in the two additional studies reviewed.

INTERPRETATION OF DIVERSITY EFFECTS

Our review shows a general tendency toward increased ecosystem functioning—and the absence of contrasting evidence of reduced ecosystem functioning—with increasing plant diversity (tables 6.1A–C). However, there are several more or less subtle differences in the shape of the biodiversity–ecosystem functioning relationship between studies. As suggested by Allison (1999) and Schläpfer and Schmid (1999) this could largely be the result of differences in the design of studies that favor one process or another that is thought to explain the particular shape of the relationship (figure 6.2). Obviously, several of these processes may operate together, and without further experimentation it is impossible to separate the processes based on the resulting overall relationships alone (figure 6.2B, left-hand side). Nevertheless, it is possible under appropriate circumstances to reject the hypothesis that one process alone provides a sufficient explanation, or the hypothesis that a particular process does not contribute at all to the observed relationship (table 6.2).

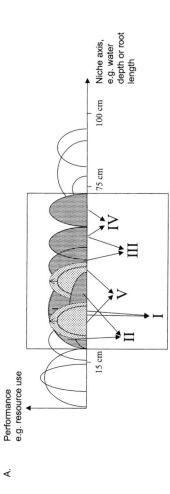

FIGURE 6.2. Illustration of processes leading to observed biodiversity–ecosystem functioning relationships (see text for details). (A) Plant species performance curves along a single niche axis to illustrate potential niche separation and overlap (ideally, the niche axis should be replaced by multidimensional niche space). The rectangle represents a section of the niche axis represented in a particular biotope. (B) Each row I–V shows from right to left: the performance curves of two selected species (in V, before and after interaction); how the two species contribute to community biomass or productivity in a competitive replacement experiment with constant planting density of 1000 individuals per square meter (where the monoculture yields are scaled to 1 and RYT = relative yield total; see, e.g., Harper 1977); the name of the process; the biodiversity–ecosystem functioning relationship that would result from the process if one to several species were assembled according to the rules illustrated by the diagram on the right. On the very left, the relationship expected from a mixture of processes is illustrated.

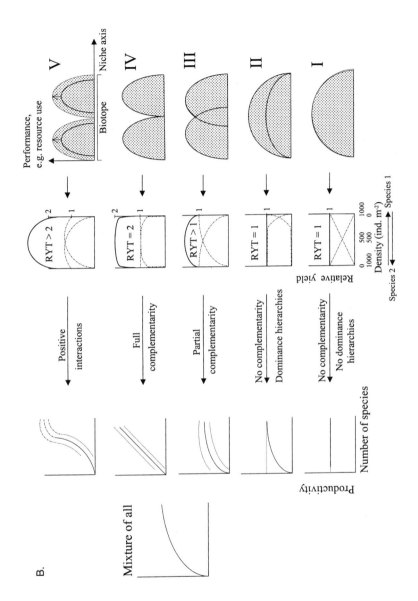

B.

TABLE 6.2. Processes for Observed Biodiversity–Ecosystem Functioning Relationships

Diversity Gradient and Study Design	Observed Ecosystem Process	General Tendency of Effects††	Number Effects		Identity Effects	Examples (studies)
			No-complementarity or Sampling Processes[a]	Complementarity or Mutualistic Processes[a]	Specific Species or Specific Species Combinations[a]	
Experimental random mixtures (long-term sowing or planting exp.)	productivity, biomass	↑, log-linear	+	+[b]	+	Naeem et al. 1996, Hector et al. 1999, Tilman, Wedin & Knops 1996, Tilman et al. 1997, Hooper & Vitousek 1998, Symstad et al. 1998
Experimental random mixtures (short-term planting exp.)	productivity, biomass	none	–	...	+	Garnier et al. 1997
Experimental nested mixtures	productivity, biomass	↑	+	...	+	Naeem et al. 1994/95, Niklaus et al. 2001b
Experimental random mixture of dead plant material	decomposition	id.	–	–	+	Wardle and Nicholson 1996, Wardle et al. 1997
Experimental mixtures, evaluation based on "relative land output" (Jolliffe 1997)	productivity, biomass	↑, weak	–	+[b]	+	Studies reviewed in Jolliffe 1997

Notes: See Table 6.1A for key.

[a] + = evidence for process; – = evidence against process; ... : = no evidence.

[b] Evidence from overyielding mixtures.

Study designs varying plant diversity, see Tables 6.1A–C.

For convenience, we assume the potential contribution of a species to ecosystem functioning can be measured as its performance along an axis representing the entire niche space available to the plant community in its biotope (figure 6.2A; see, e.g., Whittaker 1975; Hutchinson 1978). We further assume that performance is directly related to competitive ability. Without this restrictive assumption any relationship between biodiversity and ecosystem functioning can easily be produced in theory (see, e.g., Loreau 1998a). In the simplest case, all species in a community have identical performance curves (figure 6.2B, I) and persist only due to processes of random colonization and mortality (Hubbell 1979). This "no-complementarity-no-dominance" process corresponds to a situation described by Aarssen (1983) as competitive combining ability. The "sampling" process (Aarssen 1997; Huston 1997; Tilman 1997a) differs from the no-complementarity-no-dominance process in that high productivity of species assemblages is explained by the more likely occurrence of a highly productive species (figure 6.2B, II). In the purest form of the sampling process, all assemblages will become monocultures of their highest-performing species. For the BIODEPTH experiments, actual species numbers were only 10% less than starting species numbers (Hector et al. 1999) and equally productive assemblages of different initial species composition were not dominated by a few or a single species they all had in common (e.g., Joshi et al., unpublished data of the Swiss BIODEPTH site). Thus, if the first two processes were not both rejected due to the occurrence of overyielding mixtures (see Hector, chapter 4, this volume), the BIODEPTH results would have better supported the no-complementarity-no-dominance process than the sampling process. Taken together, the two processes often seem to explain part but not all of the diversity effects (table 6.2).

If the performance curves of species do not overlap completely, highly diverse species mixtures may outperform mix-

A. Random species sampling

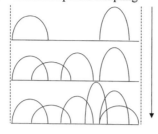

Increasing species number
Decreasing difference
Non-expanding niche axes

B. Functional group sampling

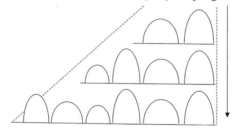

Increasing species number
Constant difference
Expanding niche axes

FIGURE 6.3. Two possibilities of sampling species from a pool (see figure 6.2). In (A), species are randomly sampled from a constant pool, thus leading to decreasing average differences between species with increasing species number. In (B), average differences between species are held constant, which requires an increasingly "broad" (in terms of covered niche axes lengths) species pool as species number is increased.

tures with lower diversity and these may again outperform monocultures. The "partial complementarity" process occurs if the performance curves still partly overlap (figure 6.2B, III) and the "full complementarity" process occurs when there is no overlap (figure 6.2B, IV). As communities are assembled in biodiversity experiments from local species pools, an increasing species number is probably always unintentionally correlated with decreasing species difference (figure 6.3A). This obviously increases the chances of demonstrating effects that are consistent with the partial complementarity process. The "positive interactions" process oc-

134

curs if the performance curves of species in a community are positively affected by biological interactions (figure 6.2B, V). At high diversities, the previously discussed processes will very likely override the effects of positive interactions. Thus the resulting relationship would turn into a logistic curve (see, e.g., Schläpfer, Schmid, and Seidl 1999). Direct positive interactions are rare among plants (Harper 1977; but see Hacker and Gaines 1997). Indirect positive interaction effects, however, may occur via species of other trophic levels such as soil microorganisms (cf. Balser, Kinzig, and Firestone, chapter 12, Naeem, chapter 5, this volume). Nevertheless, in the studies reviewed in tables 6.1A–C no logistic relationships were observed. Taken together, the occurrence of complementary or of positive interactions can be demonstrated by comparison of particular mixtures with their nested subsets and monocultures (analysis of relative land output in Jolliffe 1997, analysis of relative yield totals as indicated in figure 6.2B, and analysis of overyielding as used by Hector in chapter 4 herein).

If replication within diversity levels allows contrasts between communities in which a particular species or species combination is present or absent, it is possible to test for the occurrence of identity rather than pure number effects. We call this the *identity process.* We include idiosyncratic effects in the identity process even though it is conceivable that idiosyncratic effects are sometimes pure number effects— for example, if four-species assemblages independently of their particular composition always have particularly low ecosystem performance. Indeed, this was encountered in some situations in the BIODEPTH experiments (Joshi, unpubl. data) and in modeling studies (Kinzig and Pacala, chapter 9 herein). It is, of course, reasonable to expect that, theoretically, all pure number effects could be represented by a large set of identity effects if perfect information about all performance curves and interactions of species were available (see Symstad et al. 1998).

General Patterns under Variable Conditions

REVIEW OF EMPIRICAL STUDIES

Empirical evidence for the effects of plant diversity on the response of ecosystem functioning to perturbations comes from observational, semi-experimental (e.g., comparison of successional stages that differ in species number), and experimental studies (table 6.3). Although the available observational and semi-experimental studies should be interpreted as much in light of other explanatory variables ("third variables"), we only discuss them here with regard to diversity. In most cases, perturbations occurred naturally, and in the majority of cases we know of, the response of ecosystem performance was measured in relation to net primary productivity. Following Schläpfer and Schmid (1999) we classified the responses into resistance (smaller changes in ecosystem performance caused by perturbation indicate more resistance), resilience (degree to which preperturbation levels of ecosystem performance were regained a given time after perturbation), and invariability (inverse of some measure of temporal variability of ecosystem performance when no particular perturbation event was identified; see Tilman and Lehman, chapter 2; Chesson, Pacala, and Neuhauser, chapter 10, this volume). More diverse plant communities had greater resistance to perturbation than less diverse communities in all but four cases, one of which showed the opposite pattern (Hurd and Wolf 1974). In contrast to resistance, there was no clear pattern for the effects of plant diversity on resilience. So far, the clearest pattern emerging from the empirical studies is the positive effect of plant diversity on invariability.

INTERPRETATION OF DIVERSITY EFFECTS

The interpretation of the relationship between biodiversity and ecosystem response to perturbation has not been developed very far in the literature. Recent models (Doak et al. 1998; Tilman 1999b; Yachi and Loreau 1999) suggest simple mechanisms for potential effects of direct diversity manipulations that only now are being carried out (e.g., by A. Pfisterer et al., unpubl. data). It seems customary that a sampling process is used to explain positive biodiversity effects on ecosystem resistance and invariability: as the biotope space is shifted by perturbation it is more likely that a more diverse community already includes some species with high performance curves under the altered conditions than does a less diverse community. This interpretation of biodiversity effects by such a sampling process under variable conditions is also called the *insurance hypothesis* (Lawton and Brown 1993; Naeem 1998). Where the shape of the diversity–resistance relationship was assessed in empirical studies it was asymptotic for species number (Tilman and Downing 1994; Tilman 1996) and linear for functional group number (Joshi, Matthies, and Schmid 2000). This could be taken as an indication of a complementarity process because niche overlap is expected to be smaller among functional groups (see figure 6.2B, III, and figure 6.3B).

From reviewing empirical studies on the effects of plant diversity on ecosystem functioning it is not obvious if relationships are more strongly positive under spatially variable conditions than under uniform ones. Also, the measurements of invariability in the reported studies were never related to different levels of temporally variable conditions. Both spatial and temporal heterogeneity of an appropriate scale and predictability may allow more species to coexist and to efficiently use the available biotope space within an ecosystem than would be possible under uniform conditions (see Kinzig and Pacala, chapter 9; Chesson, Pacala, and Neuhauser, chapter 10, this volume).

TABLE 6.3. Effects of Plant Diversity on Stability Properties of Plant Community Biomass or Plant Productivity during and after Natural or Experimental Perturbations

Reference	Diversity Gradient*	Species Comp.†	Climatic Zone‡	Ecosystem Type	Perturbation‖‖
Joshi, Matthies & Schmid 2000	exp., F	rr	temp	grassland	e; +Hp
A. Pfisterer et al. (unpubl.)	exp., F	rr	temp	grassland	e; −R
	exp., F	rr	temp	grassland	e; −R
	exp., F	rr	temp	grassland	e; +Hi
B. Schmid et al. (unpubl.)	exp., F	rr	temp	grassland	e; +T
Dodd et al. 1994	nutr., F	nr	temp	grassland	n
Tilman & Downing 1994	nutr., F	nr	temp	grassland	n; −R
	nutr., F	nr	temp	grassland	n; −R
Tilman 1996	nutr., F	nr	temp	grassland	n; −R
	nutr., F	nr	temp	grassland	n; −R
Hurd & Wolf 1974	succ., F	nr	temp	old field	e; +N
Mellinger & McNaughton 1975	succ., F	nr	temp	old field	e; +N
Leps, Osbornova-Kosinova & Rejmanek 1982	succ., F	nr	temp	grassland	n; −R
	succ., F	nr	temp	grassland	n; −R
Brown & Ewel 1987	succ., F	nr	trop	plantation, forest succ.	n
Berish & Ewel 1988	succ., F	nr	trop	plantation, forest succ.	n; −R
McNaughton 1985	obs., F	nr	trop	grassland	e/n; +Hg
	obs., F	nr	trop	grassland	e; −P
	obs., F	nr	trop	grassland	n

Notes: See Table 6.1A and 6.1B for keys.
‖‖ n = natural perturbation; e = experimental perturbation; −H = herbivore exclusion; +Hg = addition of grazers; +Hi = addition of an insect herbivore; +Hp = addition of a hemiparasite; +N = increased N-supply; −P = removal of plant species; +R = increased precipitation; −R = drought; +T = trampling. With experimental perturbations, time scale refers to the duration of the perturbation, with natural perturbations to the duration of the study.

TABLE 6.3. *Continued*

Time Scale	Plot Size (m²)	Diversity Levels†	Stability Property	Specifications to Stability Property	Type of Diversity Effect††
4 m	0.25	1–32	resistance	r. to loss of aboveground biomass	↑ linear with funct. group no.
8 w	1	1–32	resistance	r. of aboveground biomasss	↓ log-linear
8 w	1	1–32	resilience	deviation 1 y after drought from controls	↓ log-linear
2 w	0.09	1–32	resistance	r. to loss of aboveground biomass	↑ log-linear
2y	4	1–32	resistance	r. of aboveground biomass	none
42 y	1000–2000	8–45	invariability	—	↑ (tendency)
2y	16	1–26	resilience	deviation 4 y after drought from pre-d. biomass	optimum
2y	16	1–26	resistance	decrease of biomass in drought rel. to normal year	↑ log-linear
2y	16	1–26	resistance	decrease of biomass in drought rel. to normal y	↑ log-linear
8 y	16	1–26	invariability	i. in non-drought ys	↑ linear
1 y	1500	~35/~50	resistance	r. to N-pulse	↓
1 y	1500	~35/~50	resistance	r. to N-pulse	↑
4 y	no inf	4–20	resistance	comparison with pre- and post-drought y	↑
2 y	no inf	4–20	resilience	prop. return during 2 y following drought	↓
2 y	256	1,40, 50,60§§	invariability	i. of herbivory	↑
1 y	256	1,40, 50,60§§	resistance	r. of fine-root biomass	none
1 m	—	~4–14	resilience	—	↑
<1 y	—	~4–14	resilience	—	none
1 y	–	~4–14	invariability	1/CV of aboveground biomass	↑

BIODIVERSITY EFFECTS AMONG TROPHIC LEVELS

Review of Empirical Studies

Studies about the effects of varying biodiversity at trophic levels other than that of the primary producers are of interest (see cases listed in table 6.4), because they show how ecosystem processes can be modified by trophic interactions. The diversity manipulations in these studies often involved several trophic levels simultaneously, as is likely to occur in species-loss scenarios, and ecosystem performance was most often measured in relation to biomass build-up (productivity) or reduction (respiration, decomposition) (table 6.4). In contrast to the plant studies reviewed before, grassland systems are greatly underrepresented in table 6.4. These studies were mostly conducted with soil or aquatic communities and in small volumes. The only pattern that emerges from these experiments so far is that, in contrast to the studies where only plant diversity is varied, there is no general tendency of positive biodiversity–ecosystem functioning relationships. However, some studies, which may be less directly relevant to whole-ecosystem models (and thus are not listed in table 6.4), report significant effects of varying biodiversity within a single trophic level other than that of the primary producers (van der Heijden et al. 1998; Norberg 2000), or of effects of plant diversity on processes at the trophic level of consumers (Mulder et al. 1999). Hence, it may be speculated, for instance, whether the general tendency for positive relationships between plant diversity and primary productivity does result in an increasing trend in herbivory and herbivore biomass in terrestrial ecosystems (McNaughton et al. 1989; Cyr and Pace 1993; Cebrian 1999).

Importance of Biological Interactions

Manipulating the diversity at a single trophic level in a multitrophic community sometimes had a stronger effect on a specific variable than manipulating the diversity of several trophic levels at once (Naeem and Li 1997; Laakso and Setälä 1999). This is an indication that, in some of the reported studies, significant but counteracting effects of diversity at single trophic levels may add up to net effects that are small or absent. Together with a strong effect of single species (Mikola and Setälä 1998; Laakso and Setälä 1999), this could explain most of the results presented in table 6.4. With regard to the primary producers a keystone species or functional group in consumer or decomposer food chains may either change the biotope space or the performance curves of plants (see figure 6.2). The first type of species may be called *ecosystem engineers* (Jones, Lawton, and Shachak 1997) or *niche-axes modifiers* and the second type, *trait-distribution modifiers*. This second term is proposed here because a change in performance curves should be related to a change in species traits that is related to resource utilization. Further, we suggest below that replacing species number or diversity by trait distributions may lead to increased predictability of ecosystem functioning.

DESIGNING EMPIRICAL STUDIES TO MEASURE BIODIVERSITY–ECOSYSTEM FUNCTIONING RELATIONSHIPS

Relevance of Existing Studies

Controversies about the relationship between biodiversity and ecosystem functioning partly arose because of some contrasting results from observational studies and experiments with model plant communities. Whereas net primary productivity is usually positively related to plant diversity in experiments (see table 6.1A), in observational studies that did not control for among-site edaphic or climatic variation, the most productive plant communities were often found to be less diverse than communities of intermediate productivity (see, e.g., Grime 1973; Huston 1979; Rosenzweig and

141

TABLE 6.4. Ecosystem Effects of Varying Biodiversity at Multiple
Trophic Levels

Reference	Diversity Gradient*	Species Comp.†	Climatic Zone‡	Ecosystem Type	Time Scale	Plot Size
Naeem et al. 1994/95	exp., P	n	temp.	ruderal	206 d	1 m²
	exp., P	n	temp.	ruderal	206 d	1 m²
	exp., P	n	temp.	ruderal	206 d	1 m²
	exp., P	n	temp.	ruderal	206 d	1 m²
	exp., P	n	temp.	ruderal	206 d	1 m²
	exp., P	n	temp.	ruderal	160 d	1 m²
Naeem & Li 1997	exp., M	r	—	aquatic	57 d	50 ml
	exp., M	r	—	aquatic	57 d	50 ml
McGrady-Steed, Harris & Morin 1997	exp., M	rr	—	aquatic	42 d	100 ml
	exp., M	rr	—	aquatic	42 d	100 ml
Mikola & Setälä 1998	exp., M	r	temp.	forest	140 d	20 ml
	exp., M	r	temp.	forest	140 d	20 ml
	exp., M	r	temp.	forest	140 d	20 ml
	exp., M	r	temp.	forest	140 d	20 ml
	exp., M	r	temp.	forest	140 d	20 ml
Laakso & Setälä 1999	exp., P	n	temp.	forest	276 d	1.5 l
	exp., P	n	temp.	forest	276 d	1.5 l
	exp., P	n	temp.	forest	276 d	1.5 l
	exp., P	n	temp.	forest	276 d	1.5 l
	exp., P	n	temp.	forest	276 d	1.5 l
Petchey et al. 1999	exp., M	n	—	aquatic	7 w	100 ml
Hurd & Wolf 1974	succ. F	nr	temp.	old-field	6 m	1500 m²
Smedes & Hurd 1981	succ. F	nr	temp.	subtidal	2 y	0.01 m²
	succ. F	nr	temp.	subtidal	2 y	0.01 m²

Notes: See Table 6.1A for key.
producer, p = autotroph (primary producer); c1 = primary consumer; c2 = secondary consumer; dec. = decomposer; f = fungivores; b = bacterivores.
*** Number of trophic groups varied.

TABLE 6.4. *Continued*

Groups with Δ Diversity##	Diversity Levels§	Troph. Gr.***	Level of Effect##	Observed Variable	Specifications to Observed Variable
p/c1/c2/ dec.	2,5,16/ 3-5/1-2/3-8	4	producer	productivity	—
p/c1/c2/ dec.	2,5,16/ 3-5/1-2/3-8	4	dec.	decomposition	short-term
p/c1/c2/ dec.	2,5,16/ 3-5/1-2/3-8	4	dec.	decomposition	longer-term
p/c1/c2/ dec.	2,5,16/ 3-5/1-2/3-8	4	several	respiration	community resp.
p/c1/c2/ dec.	2,5,16/ 3-5/1-2/3-8	4	abiotic	nutrient retention	total N
p/c1/c2/ dec.	2,5,16/ 3-5/1-2/3-8	4	abiotic	water retention	—
p/c1/c2	1-3/1-3/1-3	5	producer	biomass	
p/c1/c2	1-3/1-3/1-3	5	dec.	biomass	number of bact. cells
p/c1/c2	3-31	4	several	biomass	CO_2 production
p/c1/c2	3-31	4	dec.	biomass	org. matter dec.
c1 (b, f)	1-3/1-3	2	c1	biomass	nematode biomass
c1 (b, f)	1-3/1-3	2	dec.	biomass	microbial biomass
c1 (b, f)	1-3/1-3	2	dec.	respiration	microbial r.
c1 (b, f)	1-3/1-3	2	dec.	C and N min.	—
c1 (b, f)	1-3/1-3	2	c2	biomass	—
c1/dec.	1,5/1,5	2	producer	biomass	—
c1/dec.	1,5/1,5	2	c1	biomass	—
c1/dec.	1,5/1,5	2	several	respiration	—
c1/dec.	1,5/1,5	2	abiotic	soil pH	—
c1/dec.	1,5/1,5	2	abiotic	water content	—
p/c1/c2/b	1-5/1,3/ 0-3/1,3	4	all	biomass	—
p, c1	~35, ~50 (p)	2	c2	resistance	r. to N-pulse pert.
c1, c2, dec.	~30, ~35 (total)	3	several	resistance	r. to predation
c1, c2, dec.	~30, ~35 (total)	3	several	resilience	r. after predation

143

Abramsky 1993; Grace 1999; Waide et al. 1999). It may therefore be argued that experiments are not realistic because they exclude this among-site abiotic variation and only vary diversity, which in nature never occurs. However, experiments do not intend to mimic the situation typical for observational studies, rather they try to predict what would happen if species go extinct in the future due to human influence, a situation that obviously cannot be analyzed in short-term observational studies. Nevertheless, the relevance of observational studies could be increased if researchers selected sites of similar abiotic conditions or if they accounted for among-site abiotic variation such as soil fertility or successional age by including these "third variables" as covariates in statistical analysis.

The relevance of the experimental studies done so far can be assessed in two ways: one relating to the kind of differences among species making up the experimental communities and the other relating to the procedure by which the diversity treatments are established.

As can be seen from modeling (e.g., in Kinzig and Pacala, chapter 9; Chesson, Pacala, and Neuhauser, chapter 10; Holt and Loreau, chapter 11, this volume; see also Ives, Gross, and Klug 1999), the effect of biodiversity on ecosystem functioning actually depends on the ecological differences among the species within a community, for which the number of species is one simple measure. However, ecological differences are not easily measured because many traits may affect resource utilization and plant performance along the niche axes represented in the biotope space (see figure 6.2a). We therefore suggest that differences be measured by numbers of functional groups as an approximation to multivariate trait distribution of a community with functional groups defined according to a set of ecologically relevant characteristics using objective methods (Pillar 1999; Sullivan and Zedler 1999; Weiher et al. 1999). In experimental designs the number of functional groups may be increased

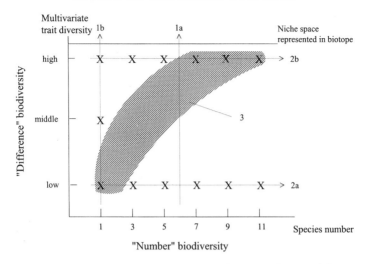

FIGURE 6.4. "Species number x difference" design plane to define treatments. Designs from vertical slices keep the number of species constant but increase the number of functional groups (1a) or the phenotypic or genetic variability of species (1b). Designs from horizontal slices keep the number of functional groups constant but increase the number of species (2a: experiments with narrow-niched species; 2b: experiments with broad-niched species). Most of the empirical studies reviewed in this chapter implicitly used a design lying within the shaded area (3; see text for explanation).

while the number of species is held constant (vertical slices in figure 6.4). In the BIODEPTH experiments, for instance, this was the case for communities consisting of four or eight species (Hector et al. 1999). In the extreme case where communities consist of single species, functional diversity may still be varied by increasing phenotypic or genetic variability, which directly relates to the question of population viability (Fischer and Matthies 1998). Alternatively, in experimental designs the number of species may be increased while the number of functional groups is held constant (horizontal slices in figure 6.4). In the BIODEPTH experiments, for example, this was the case for communities consisting of grasses only (Hector et al. 1999). While the upper end for the number of species is only limited by the species pool, the upper limit of the number of functional groups used in an experiment should ideally be large enough to cover the entire niche represented in the biotope space. This assumes that species from outside this range would not be able to establish in the community (not pass the site filter shown in figure 6.1) and does not consider the possibility that invading species could expand niche axes. From this it becomes clear that as the number of species is increased further and further, the functional group or trait diversity must reach this upper bound, and therefore the number of species inevitably becomes negatively correlated with the average pairwise ecological difference between species. It seems that this occurred in the majority of experimental studies reviewed here and is consistent with the curvilinear relationships often found between biodiversity—measured by species number—and ecosystem functioning.

The procedure by which the diversity treatments are established influences the relevance of experimental studies to natural ecosystems and to those simplified by humans. We have illustrated these processes in figure 6.1, and Holt and Loreau discuss them in chapter 11 of this book. The natural assembly of a community from the large global species pool can be understood as a several-stage filter process (Diaz, Cabido, and Casanoves 1999). In natural communities, species composition (and number) is determined by a "biogeographic filter," a "local dispersal filter," and a "site filter." Depending on the specific filter, the diversity that passes it may or may not be a random sample with respect to species functional traits. In other words, reduction of species number by a filter does not necessarily imply an unbiased reduction of functional trait diversity. However, experimental studies do not explicitly state which filters they intend to mimic. For example, Ewel, Mazzarino, and Berish (1991) compared a naturally assembling successional community with an experimentally enriched successional community, thus manipulating the dispersal filter between regional and local species pools. Other experimental studies take the 20 to 50 or so species of a local community as their regional species pool. This pool, if it excludes species that are only represented by propagules in the local community, can be even more restricted than the (natural) local species pool. Random or other types of sampling from the experimental regional pool are then used to create different experimental local pools. These experimental local pools are taken as the diversity treatments. If all of the species do not pass the subsequent site filter, for example, due to competitive exclusion, and if abundance distributions in the resulting communities are uneven, this leads to effects that can be explained by the sampling process. By simulating different local species pools and subsequently observing actual species numbers, experimental studies might demonstrate how specific site filters, including management practices, operate and thus influence community assembly and ecosystem functioning. Indeed, one experimental study deliberately manipulated a site filter by adding nutrients to establish diversity treatments (Tilman and Downing 1994).

In a way, the situation in observational studies is opposite to many experimental studies: here the actual species number is used as the diversity variable and the diversity of the local species pool is not known. In summary, experimental studies have focused on "pure" biodiversity effects resulting from manipulating species pools (biogeographic and local dispersal filters), equivalent to simulating arbitrary or random species extinction (e.g., due to overharvesting or due to habitat destruction and fragmentation; see also chapter 13 herein, Lawler, Armesto, and Kareiva). Observational studies, however, have focused on biodiversity effects resulting from differential community assembly due to the action of different site filters (figure 6.1). Often, natural community assembly may not be random with respect to the relative effects of species on ecosystem functioning (Grime 1998;

Wardle 1999; but cf. Gleason 1926), while the natural assembly of species pools may be more likely to be so.

Suggestions for Future Studies

As expected, our review of biodiversity studies provides limited empirical information for use in ecosystem models. Gaps exist because, to date, only certain aspects of diversity have been treated and only certain ecosystem variables have been measured. However, the information provided by empirical studies is not only limited because of omissions but also because of the lack of explicit descriptions of diversity treatments and sometimes of measurements. There are three important aspects of biodiversity: numbers, differences, and abundances of species. Often experimental studies are unclear about the difference and abundance aspects, that is, they assume differences to be constant and abundances to be even. To improve this situation we suggest that future empirical studies should

- try to use ecological trait diversity instead of or along with species diversity,
- always report achieved abundances together with species numbers (e.g., effective diversity, evenness, or dominance structure).

Future observational studies should also make sure that edaphic, climatic, or disturbance differences among units of observation are eliminated either by careful selection of sites or by statistical analysis. About measurements, the most important recommendation for future experiments, and even more so for future observational studies, is the inclusion of further variables in addition to aboveground net primary productivity. These further variables are measurements of belowground processes related to mineralization and decomposition and measurements of plant cover or leaf-area index related to transpiration.

Many problems may make it difficult to apply experi-

148

ments directly to natural or human-dominated ecosystems. These problems of experimental studies include their short-term duration, the even-aged community composition, artificial maintenance of communities by repeated seeding of wanted species, and the "unnatural" simulation of species extinction. To exclude unwanted sampling effects (but retain those effective in natural systems) one could design two-stage experiments. This approach corresponds to the modeling studies of Loreau and Mouquet (1999) and of Holt and Loreau, chapter 11, this volume. In the first stage, communities would be allowed to assemble from simulated local pools of varied species number. In the second stage, only those diversity treatments would be continued where input and effective diversities did not diverge (i.e., species number in seed mixture ≅ species number in community). This would avoid the problem that communities with the same actual diversities are derived from differently sized species pools. That these communities no longer have fully randomized species compositions need not be a problem. Another type of two-stage experiment would start with the same first stage, but in the second stage, only communities of the same actual diversities derived from different input diversities would be used. This would allow proper analysis of the sampling process.

Addressing the effects of specific nonrandom extinction processes could be achieved by the following two-stage setup. In the first stage, communities of different species richness would be generated from a regional species pool by introducing different realistic anthropogenic species filters such as nutrient enrichment. Then the resulting species mixtures would be replanted under identical conditions (not-enriched soil) until they reach equilibrium abundance distributions. Plotting the productivities of the resulting second-stage communities against the diversities in the first-stage communities would yield the effect of species reductions that are specific to a particular extinction process such as, in this case, a temporary nutrient enrichment. This could then be compared to effects from random species extinction.

The above recommendations refer to studies under uniform conditions. To assess effects of biodiversity on ecosystem responses under variable conditions it is convenient to use natural or experimental perturbations. The latter can mimic events that are predicted to increase in severity or frequency due to global change. Such experiments have only recently been started and should be done more often in the future. Another urgent need is the study of effects of environmental heterogeneity on the strength of the bio-diversity–ecosystem functioning relationship. Differences in the variation and average levels of performance of intact and depauperate model ecosystems may be greater with appropriate levels of spatial and temporal heterogeneity than under more stable environmental conditions. Our hypothesis is that, for example, the amount of overyielding in plant communities (see Hector, chapter 4 herein) is positively correlated with increasing environmental heterogeneity of this kind. To our knowledge, neither empirical nor modeling studies have ever addressed such a hypothesis.

ACKNOWLEDGMENTS

We thank A. Hector, A. Kinzig, and D. Tilman for very helpful suggestions to improve the manuscript, and E. Spehn and M. Diemer for help in interpreting studies reporting ecophysiological measurements.

150

CHAPTER SEVEN

The Transition from Sampling to Complementarity

Stephen Pacala and David Tilman

We suspect that only two years ago many ecologists would have predicted that the sampling mechanism (Huston 1997; Aarssen 1997; Tilman et al. 1997) ultimately would be confirmed as the primary cause of the positive effect of biodiversity on ecosystem function reported in most experimental studies. Indeed this view catalyzed a largely semantic debate among scientists about the interpretation of the experimental findings. One camp (Huston 1997; Aarssen 1997; Wardle, Bonner, and Nicholson 1997; Grime 1997) argued that the experiments showed no "real" effect of diversity because the most productive monocultures performed as well in the experiments as the high-diversity plots. The other camp (Tilman et al. 1997; Naeem et al. 1995; Lawton et al. 1998) argued that an essential feature of diversity is that it increases the probability that species with superior traits will be present and that a variety of processes in nature operate approximately like the sampling mechanism. These include anthropogenic species removals and natural biogeographic colonization and extinction, especially in a variable environment. Thus, the greater the number of species that colonize a vacant site, the greater the probability that at least one would be well adapted to the local conditions, and the greater the number of species lost from a habitat, the greater the probability that one or more of these would have been major contributors to ecosystem functioning.

But now, only two years later, this issue has been displaced by the unanticipated results from the largest and longest-running experiments. Three general findings emerge from chapters 2–4 of this volume and from Tilman et al. (2001).

1. Most experimental studies initially produce a positive relationship between diversity and function that looks like the work of the sampling mechanism—an increasing mean but a triangular pattern of scatter with a flat top.
2. After a small number of years, a transition occurs to a pattern of overyielding that looks like the signature of niche complementarity, with the best high-diversity plots significantly outperforming the best monocultures.
3. Conspicuous exceptions occur in a minority of experiments, such as inverse relationships between diversity and function in Hooper and Vitousek (1997) and in the BioDep experiments in Greece (but not in the other BioDep locations, Hector et al. 1999).

Of these, the transition from the sampling mechanism to the niche is surely the most intriguing. Why do so many experimental systems behave initially like the simple sampling model? Why does overyielding emerge, but only after an extended time lag? What awakens MacArthur's ghost?

We suggest that this kind of transition is precisely what should occur because of a basic property of nonlinear population dynamics and of plant growth in monoculture. Specifically, we suggest that the initial dynamics of communities are transients driven mainly by interspecific differences in maximal growth rates, whereas the longer-term dynamics are driven more by interspecific competition. As emphasized in Huston (1997), the sampling effect can result from the greater probability that a species with a high rate of vegetative growth would be present at increased diversity. If such transient dynamics are eventually overcome by interspecific competition resulting from niche differentiation, then the initial effects of diversity would be sampling effects and the long-term effects would be niche effects. This possibility holds the promise both of explaining the recent experimental results, and of resolving the debate over the importance of sampling versus niche effects.

Consider first the simplest of ecological models of competition: the Lotka-Volterra (LV) competition equations. The abundance $(B_i(t))$ of a single LV species growing alone is given by the solution of the logistic equation:

$$B_i(t) = \frac{K_i}{1 + \left(\dfrac{K_i}{b_0} - 1\right)e^{-r_i t}}$$

where b_0 is the initial abundance. If the initial abundance is well beneath carrying capacity, then $B_i(t)$ will initially increase exponentially at rate r_i. Slightly less well known is that growth also slows down exponentially at this same rate, r_i, as abundance nears carrying capacity (deviations of $B(t)$ from $K(t)$ shrink as e^{-rt}).

Now consider what happens initially if we grow two such species together. We suppose that we start at equal abundance well beneath carrying capacity, and set the measurement scale so that the initial abundances equal 1. Because each species grows exponentially, the species with the largest intrinsic rate of increase pulls ahead of the other exponentially (the ratio $B_1(t)/B_2(t)$ initially equals approximately $e^{(r1-r2)t}$). Thus, in a graph of B_1 versus B_2 the initial trajectory of the abundances approximately follows the path:

$$B_1(t) \propto B_2(t)^{r_1/r_2}.$$

Suppose that we label species-1 as the species with the larger initial growth rate. If either r_1 is significantly larger than r_2 or the species start well below carrying capacity, or some combination of the two, then the initial trajectory will be

LV Competition

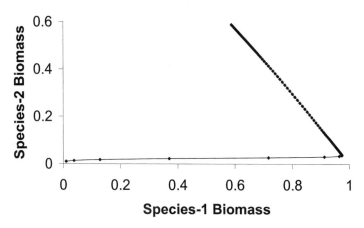

FIGURE 7.1. A trajectory from the Lotka-Volterra competition equations. The trajectory starts near the origin and the dots show equal time intervals. Parameters are $K_1 = K_2 = 1$, $r_1 = 2.5$, $r_2 = 0.5$, $\alpha_{12} = \alpha_{21} = 0.7$.

nearly horizontal, like that shown in figure 7.1 (trajectory starts near the origin and dots are placed at equal time intervals). The faster-growing species will sprint ahead of the slower-growing species and then slow down quickly (at rate $-r_1$) as it nears its carrying capacity.

The upshot of this period of initial growth is that the total biomass of the two-species community will be approximately equal to K_1, which is also approximately the biomass attained by a monoculture of the fast-growing species over the same time interval. It is trivial to show that this result also extends to the initial behavior of communities with more than two species and to other models of competition, because an initial period of exponential growth is a general property of population dynamic models (at least for internally recruiting populations).

Plants in biodiversity–ecosystem function experiments typically have both characteristics that lead to initial takeover by the fastest-growing species. The experiments start at low initial biomass; for example, total plant biomass immediately after germination was approximately 1000-fold less than final values in Cedar Creek experiments. Also, the experiments inevitably contain a mix of species well- and poorly adapted to local conditions as well as a mix of successional types. Each of these leads to a wide range of r's.

Returning to the simple LV example, we left our two species with the faster-growing species-1 near its carrying capacity and the slower-growing species-2 at relatively low abundance. Subsequent dynamics will have a qualitatively different character, because they are now dominated by competition and density dependence, rather than by simple exponential growth. If, in addition to having smaller r, species-2 is a poorer competitor than species-1, then species-2 will simply be driven extinct. However, if species-2 is able to coexist with or displace species-1 (if $K_2 > \alpha_{21}K_1$), then the trajectory of $(B_1(t), B_2(t))$ will switch direction. Rather than continue to grow horizontally, it will begin to move diagonally up and to the left, but at a much slower rate than the precompetitive dynamics (figure 7.1; note much closer spacing of equal-time dots along the diagonal trajectory toward the two-species equilibrium point). If $r_2 << r_1$, then the biomasses of the two species will move approximately along the isocline (or eigenvector): $B_2(t) = K_1/\alpha_{12} - B_1(t)/\alpha_{12}$, at the relatively slow exponential rate (or eigenvalue): r_2 $(1 - \alpha_{21}K_1/K_2)$. Note that this rate is slower even than r_2. The slow competitive phase of dynamics will continue until the biomasses reach whichever equilibrium is stable (either $(0,K_2)$ or the coexistence equilibrium). One final point is that the condition for coexistence ($K_1 > \alpha_{12}K_2$ and $K_2 > \alpha_{21}$ K_1) is also a necessary condition for the equilibrium yield of the two-species community to exceed the yield of either species growing in monoculture. Overyielding is inevitable if both α's are sufficiently small, and thus occurs whenever niche differentiation is sufficiently large.

Like the period of initial exponential takeover by the faster-growing species, the occurrence of a slow competitive aftermath is a general characteristic of the dynamics of competition. Together, the exponential and competitive phases of dynamics predict two of the characteristics of the majority of biodiversity–ecosystem function experiments: (1) mixtures never outperform the best monocultures initially, but can do so after a considerable time lag, and (2) a mixture of species that can coexist will often outyield all monocultures of its members.

What is still missing from this picture is an explanation of the characteristic signature of the sampling mechanism in the initial phase of dynamics—the fact that the best mixtures initially perform approximately as well as the best monocultures. In models, the initial phase of growth ends with each mixture's yield approximately equaling the monoculture yield of the species with the most rapid initial growth rate. Thus, if $r_1 > r_2$ in the LV equations, the yield of the two-species mixture will approach K_1 at rate r_1 during the initial phase. The yield of the species-1 monoculture will also approach K_1 at rate r_1, but the yield of the species-2 monoculture will lag behind. We thus obtain the flat-topped signature of the sampling mechanism from the model as a transient, because, during the initial phase of an experiment, both the largest monoculture yield and the yield in mixture are created by the same processes: exponential growth followed by a stalling near the monoculture equilibrium of the fastest-growing species.

This transient state will persist for a considerable period of time if the species with the highest r_i also has the highest K_i. In contrast, if a slower-growing species has the largest K_i, then its monoculture yield may subsequently surpass the yield of a mixture still stalled near the carrying capacity of the fastest-growing species. Although we might expect a positive correlation between initial growth rate and monoculture equilibrium biomass in a mix of species well and poorly adapted to the site, we do not expect it in general. For example, weedy plants initially outgrow woody plants, but have much lower equilibrium biomass in monoculture. Thus, the LV equations imply that the occurrence of the sampling pattern might be very short lived in some systems.

Unfortunately, the idea that the flat top of the scatter in the sampling pattern occurs because mixtures and the fastest-growing monocultures approach the carrying capacity of the fastest-growing species is not consistent with experimental observations at Cedar Creek. In both mixtures and monocultures, biomass continued to grow at Cedar Creek over the two to three years in which results consistent with the sampling hypothesis were exhibited. We thus require an explanation that does not rely on approximate equilibration.

We now restrict the generality of our argument, but only slightly, by adding a widely reported characteristic of resource-dependent plant growth. The "Law of Constant Final Yield" has been reported in hundreds of studies (Kira, Ogawa, and Shinozake 1953; and see Harper 1977). The example in figure 7.2 is from Donald's (1951) study of *Lolium loliaceum*. This empirical relationship states that final yield in an experiment is approximately independent of sowing densities for sowing densities that are sufficiently large. Thus, the yield at 210 days in figure 7.2 is approximately 1000 g/m^2 over the range of sowing densities from 2000 to $50,000/m^2$. But notice that the relationship also is flat at 130 days from planting even though the asymptotic yield then is only one-half the value at 210 days. This result is typical of studies that report intermediate yields; initial monoculture yields become independent of sowing densities well before biomass equilibrates (see examples in Harper 1977). This observation holds the final key for explaining the initial prevalence of the sampling mechanism in

157

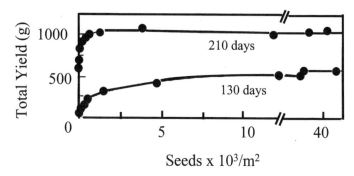

FIGURE 7.2. The Law of Constant Final Yield in *Lolium loliaceaum*, from Donald (1951), as graphed in Harper (1977, p. 153).

biodiversity–ecosystem function experiments and is not exhibited by simple "plant-free" competition models, such as the LV equations.

Let us now revisit the expected pattern of dynamics in a biodiversity–ecosystem function experiment, assuming the constant yield relationship. An initial period of exponential growth is still inevitable. Thus, mixtures will be dominated initially by the fastest-growing species and each treatment will behave like a monoculture. *Well before* these effective monocultures equilibrate, they will begin to exhibit the constant yield relationship. This is important because the design of biodiversity–ecosystem function experiments imposes a constant total planting biomass of all species put together. The initial abundance of the fastest-growing species in a 10-species plot is only one-tenth that in a single-species plot.

Without the constant yield relationship, exponential growth would not produce the flat-topped scatter of the sampling mechanism, because purely exponential growth preserves differences in initial biomasses. If growth were purely exponential, then the yield of the fastest-growing species in a Q-species mixture would decrease hyperbolically with Q as: $(b_0/Q)e^{rt}$. However, with the nonexponential mechanisms that lead to the constant yield relationship, the biomass of the fastest-growing species will approach the same value in all treatments despite differences in initial abundance. The characteristic signature of the sampling mechanism emerges early in experiments because of the combined action of exponential growth, which produces effective monocultures, and the mechanisms responsible for the constant yield relationship. These mechanisms erase differences in initial biomass while growth is still rapid and well before an effective monoculture approaches equilibrium and slower-growing species begin to gain ground.

The simple explanation of the constant yield relationship is that "the resource-supplying power of the environment comes to dominate the rate at which members of a population grow and ultimately sets the limit to yield, irrespective of plant density" (Harper 1977, p. 154). In other words, before the rate of resource capture is larger than the rate at which resources are supplied (i.e., photon flux density, precipitation rate, or N-mineralization rate), increases in biomass lead to ever-greater resource capture and exponentially compounded growth. But once biomass is large enough to capture most of the limiting resources supplied, the rate of total biomass accumulation becomes approximately constant and yield grows linearly rather than exponentially. For this reason, mechanistic models of plant competition include production functions that saturate, either directly or indirectly, at high biomass. This explains the prevalence of Michaelis Menton functions for gross production in the mechanistic models studied in later chapters.

To understand precisely why models with saturating functions explain both the constant yield relationship and the predominant results of biodiversity–ecosystem function experiments, consider the following simple example. We approximate a smooth saturating function relating biomass and production by two line segments. If biomass is less than R/r_j, then B_j grows exponentially at rate r_j. However, if

$B_j > R/r_j$, then growth saturates and the stand gains biomass at the constant rate R. The threshold R/r_j is the biomass at which potential uptake equals supply, and R is the rate of resource supply in units of equivalent plant biomass.

If initial biomass is less than R/r_j, then there are two distinct phases of growth. Production grows exponentially as $r_j b_0 \exp(r_j t)$ until it saturates at time $t^* = \ln(R/(r_j b_0))/r_j$, and then grows *linearly* as Rt. When the period of linear growth becomes sufficiently long relative to the period of exponential growth, then the total production becomes approximately independent of the initial density and we have the constant yield relationship in a growing monoculture. For $t > t^*$, total production is equal to:

$$tR + \frac{R}{r_j}\left(1 - \ln\left(\frac{R}{b_0 r_j}\right)\right).$$

To see that this depends only weakly on initial biomass b_0, suppose that $R = 0.1$ kg/m^2/month, $r_j = 2$/month, and $b_0 = 0.001$ kg/m^2. These values are appropriate for herbaceous species with NPP of approximately 500–600 g/m^2/y, canopy closure (t^*) in two months, and an initial germinating biomass of 1 g/m^2 (approximately a thousandfold less than an equilibrium biomass of 1 kg). After six months, the values predict a total production of 0.44 kg/m^2 in monoculture. If we now reduce the initial biomass of species-j by a factor of two, as it would be reduced in a two-species plot in a biodiversity–ecosystem function experiment, then its total production falls only slightly to 0.42 kg/m^2. The period of saturated linear growth largely removes the effect of initial biomass.

A simple Q-species extension of the model serves to illustrate all of the phases of a biodiversity–ecosystem function experiment. Within a season of length T, species-j biomass

160

grows at its unsaturated rate r_j if this is less than the saturated rate:

$$R\left[\frac{1-\varepsilon}{\sum_{k=1}^{Q} B_k} + \frac{\varepsilon}{B_j}\right]$$

and at the above saturated rate otherwise. The parameter ε represents the niches responsible for overyielding. When growing at the saturated rates, a fraction $(1 - \varepsilon)$ of the production R is divided among the species in proportion to their relative biomasses. Each species also receives an additional amount of production, εR, to which it alone has access. Note that total saturated production is an increasing function of ε: $R(1 - \varepsilon + Q\varepsilon)$, and that the model reduces to the single-species case that we just examined when $Q = 1$. Finally, we suppose that once the growing season is over, species-j loses a fraction v_j of its biomass to mortality, respiration, and retranslocation before the beginning of the next season.

The two-species example in figure 7.3 ($r_1 = 2.5$, $r_2 = 0.5$, $v_1 = 0.7$, $v_2 = 0.5$, $R = 0.1$, $\varepsilon = 0.5$, $b_0 = 0.001$) shows exponential takeover by the fast-growing species-1 by the end of the first year (squares show end-of-season proportion of species-1 in two-species plots). At this time, the biomass of species-1 monoculture plots (triangles) equals the biomass of two-species mixtures (circles), because the period of saturated growth eliminates the twofold difference in the initial abundance of species-1 in these plots. By the end of year 2, biomasses in the two-species plots approach the boundary equilibrium where species-1 is near its monoculture equilibrium and species-two is still near zero. The flat-topped pattern of the sampling mechanism gives way to the overyielding pattern of niche partitioning in all years after year 2. This same sequence of events also occurs in multi-

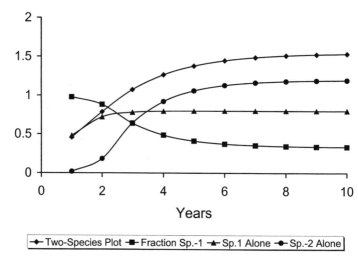

FIGURE 7.3. The transition from sampling mechanism to niche predicted by our two-species model with saturating growth. The vertical axis gives the proportion of species-1 at the end of each growing season for the curve marked by squares and end-of-season yield (kg/m^2) for the remaining curves. Parameter values are given in the text.

species experiments. The 10-species example in figure 7.4a shows the sampling mechanism at the end of the first year (r_j's spread evenly from 0.5 to 2.5, $v = 0.5$ for all species, $R = 0.3$, $\varepsilon = 0.05$, and $b_0 = 0.001$). Note that the year-1 biomass in the 10-species plot equals that of the most productive monoculture despite the 10-fold difference in the initial biomass of the fastest-growing species. In contrast, by year 3 the pattern of overyielding is well developed, with the 10-species plots having one-third more biomass than the highest-yielding monoculture (figure 7.4b).

Now, what about the exceptions to the usual pattern that occur in the Greek BioDep experiments and in the Californian grassland studied by Hooper and Visousek (1997)? It is interesting that both of these examples occur in Mediterranean climates, where herbaceous communities typically rely on winter rains, with a yearly within-season succession from

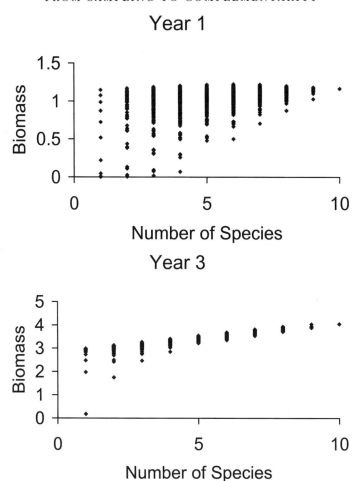

FIGURE 7.4. The transition from sampling mechanism to niche predicted by a 10-species model with saturating growth. Parameter values are given in the text.

wet-adapted species that dominate early to competitive species that dominate late (see Gulmon and Mooney 1986).

The example in figure 7.5 was constructed to give succession like that in forests ($r_1 = 1.5$, $r_2 = 0.2$, $v_1 = 0.25$, $v_2 = 0.08$, $R = 0.3$), and no overyielding ($\varepsilon = 0.0$), but it is

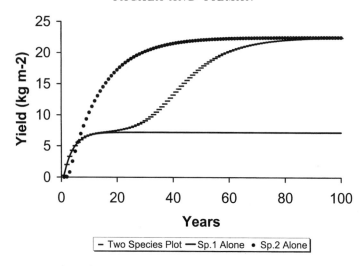

FIGURE 7.5. The transient occurrence of an inverse relationship between number of species and yield from a two-species model of successional diversity. Parameter values are given in the text.

easy to construct examples where succession is rapid. Again we see the sampling mechanism for the first several years: the biomass of two-species plots initially equals that of the fastest-growing monoculture. However, from year 10 until approximately year 50 in figure 7.5, there is an inverse relationship between number of species and biomass because the biomass in two-species plots is less than the average of the monocultures. Finally, once succession is complete, the positive relationship with the sampling pattern of scatter reasserts itself (year 100 in the figure).

To understand the transient occurrence of an inverse relationship between diversity and biomass, note that from years 10 to 20, the two-species plots are still dominated almost entirely by species-1 (mixed plots approximately equal species-1 monocultures), even though species-2 monocultures have surpassed all other stands. Evidently, competition with species-1 delays the biomass growth of species-2. Because of this competitive delay and the fact that the late-successional species has higher equilibrium abundance in monoculture than the early-successional species, a transient period occurs where late-successional monocultures outperform mixed stands. As shown by Gulmon and Mooney (1986), the presence of the low-yielding early-season dominant *Plantago* reduces the biomass of the high-yielding late-season dominant *Hemizonia* in the system studied by Hooper and Vitousek (1997).

The mechanism responsible for the inverse relationship between diversity and biomass in figure 7.5 is the same one discussed in the context of the LV competition model—a negative correlation between r_j and K_j. However, because of the saturated growth in the model leading to figure 7.5, the period of inverse relationship is sandwiched between periods in which biomass increases with diversity in the pattern typical of the sampling mechanism. Because we explain exceptions to the general pattern of a transition from sampling to niche as the result of unusually rapid succession, we expect that a focus on successional diversity in long-term experiments would uncover many other examples. Other possible explanations of an inverse relationship between diversity and biomass include allelopathy and luxury uptake of resources.

CONCLUSIONS

1. A transition from one behavior early in a biodiversity–ecosystem functioning experiment to another later on is expected because of some of the most general properties of competitive dynamics. Exponential growth dominates early, while competition dominates later on. Because transients are expected, it is vital to run experiments for many years.

2. The early phase of growth produces the pattern of the sampling mechanism, with an increasing mean and flat-topped scatter, because of a combination of exponential

and saturated growth. Initially, the species with the fastest exponential growth rate(s) quickly and exponentially dominate mixed plots. These species eventually reach a biomass at which the rate of resource interception equals the rate of resource supply. The biomass growth rate of the stand then saturates at a maximum value, causing linear rather than exponential growth. Linear growth erases the effects of differing initial abundances of the fastest-growing species in the experiment and leads to the "Law of Constant Final Yield." Once the constant yield relationship becomes established, we observe the signature of the sampling hypothesis: the biomass in each plot approximately equals the monoculture yield of the fastest-growing species it contains.

3. A transition from the sampling pattern to one of overyielding occurs after a time lag, because the slow dynamics of competition eventually replace the rapid dynamics of initial growth. This transition leads to overyielding if niche complementarity occurs. It is imperative to discover the mechanism behind the observed overyielding.

4. We hypothesize that an inverse relationship between biomass and diversity has probably been observed in some studies because of rapid within-season succession. Early-successional species (or early-season species) often have a high initial growth rate and low equilibrium biomass, while late-successional species often have the reverse. If competition with early-successional species suppresses the growth of late-successional species, then an initial phase dominated by rapid growth will be followed by a second transient phase in which monocultures of late-successional species outperform mixed stands. An inverse relationship could also occur because of antagonistic interactions in which better competitors produce less biomass than weaker competitors, presumably because they allocate more to their skills in antagonistic interactions and less to resource capture and efficiency of use.

166

PART 2

Theoretical Extensions

CHAPTER EIGHT

Introduction to Theory and the Common Ecosystem Model

Stephen Pacala and Ann P. Kinzig

Ultimately, the form of the diversity–functioning relationship in particular systems must depend on the mechanisms permitting species coexistence in those systems. Do those characteristics permitting competitive dominance also permit maximum functioning? How do those characteristics permitting coexistence in the presence of a competitive dominant contribute to functioning? What relationship between diversity and functioning, for instance, should we expect if coexistence derives from trophic interactions and the function of interest is nitrogen mineralization? Presumably, that relationship might look very different if, instead, coexistence derived from competition for multiple limiting resources. In other words, we need to be able to examine whether the characteristics that serve in the coexistence domain also serve in contributing to ecosystem functioning. Ecosystem functioning could be measured in terms of rates (e.g., N-mineralization), stocks (e.g., carbon storage), or resilience and stability (e.g., ability to withstand disturbance or recover from disturbance). A theoretical analysis is certainly warranted in examining the relationship among diversity, functioning, and coexistence mechanisms.

Furthermore, empirical analyses are, by necessity, of short time duration and conducted on small spatial scales. Extending these empirical results to the larger spatial scales and longer time durations that are properly the interest of scientists and managers also requires development of a body of theory applicable to different ecosystems, governed by different structuring mechanisms.

We therefore present three theory chapters, examining a variety of different coexistence mechanisms. Chapter 9 (Kinzig and Pacala) examines systems where coexistence is permitted either by successional niches or by competition-colonization trade-offs. Chapter 10 (Chesson, Pacala, and Neuhauser) looks at coexistence driven by resource partitioning either in space or in time. Chapter 11 (Holt and Loreau) applies to systems where trophic interactions allow species coexistence. In most of these chapters, we concentrate on rates (N-mineralization, NPP, and evapotranspiration) or stocks (total carbon storage, total living biomass) as our measures of ecosystem functioning.

We expect the diversity–functioning relationship to vary among these systems characterized by different coexistence mechanisms. However, in order to adequately distinguish the influence of coexistence mechanisms on diversity–functioning relationships from other possible variations in assumptions or approaches in the theory chapters, we have asked all authors to use a common ecosystem model. Thus, the functional forms for ecosystem processes (carbon and nitrogen cycling, evapotranspiration, etc.) are prescribed; the values these processes take in any particular system depend on species diversity, which ultimately depends on coexistence mechanisms. This common ecosystem model then serves as the base, if you will, of a coupled model of coexistence and ecosystem process. Each model of coexistence considered in the following three chapters is "attached" to this common ecosystem model; diversity is manipulated within the model of coexistence and the impacts of functioning examined through the ecosystem-process model.

We present the basic form of the common ecosystem model below. Authors in subsequent chapters may have made minor modifications to this base model—eliminating light-dependent production, for instance, in situations where light limitation is expected to be insignificant. These modifications are indicated within the relevant chapters.

THE COMMON ECOSYSTEM MODEL

The basic ecosystem model contains reservoirs for plant biomass (B), soil inorganic nitrogen (N), fast and slow pools of soil carbon (C_f and C_s), and soil water (W). The equation for plant growth is given by

$$B = BP[B,N,W] - (\gamma + \mu)B \qquad (8.1)$$

where $P\ [B,N,W]$ = gross primary production per unit of biomass (function of total biomass, soil nitrogen, and water)

γ = respiration per unit of biomass

μ = tissue death per unit of biomass

The production function assumes that light, soil nitrogen, and water can all limit growth; P is given by

$$P[B,N,W] = \left[\frac{r}{K_L + B}\right]\left[\frac{N^{\theta_N}}{N^{\theta_N} + K_N}\right]\left[\frac{W^{\theta_W}}{W^{\theta_W} + K_W}\right] \qquad (8.2)$$

The light function derives from the assumption that light is extinguished exponentially with increasing biomass and that growth follows Michaelis-Menton kinetics with respect to light availability; if $L = e^{-\alpha B}$, then $r'\ L / (L + C) = r' / (Ce^{\alpha B} + 1) \approx r' / (C + C\alpha B + 1) = r / (K_L + B)$.

The given functional form for nitrogen uptake is chosen so that growth "shuts off" (or at least becomes very small) for small N, just as it does in the widely used CENTURY model (Parton et al. 1987, 1996). For simplicity, most authors take $\theta_N = 1$.

As water availability declines, water stress should increase and stomates should be closed a greater fraction of the time. Again, for simplicity, we often take $\theta_W = 1$.

The equation governing soil–water availability is given by

$$W = R - sW - \frac{BP[B,N,W]}{U} - \varepsilon W \qquad (8.3)$$

where R = rainfall

$s W$ = percolation of water away from rooting zone

$B P/U$ = transpiration

U = water-use efficiency

εW = evaporation

This is a "one bucket" model of hydrology. Note that evaporation might actually decrease with increasing B due to shading, or that the size of the "bucket" might increase with increasing B because rooting depth increases. We will ignore those effects in our analyses.

The equation governing soil nitrogen availability is given by

$$N = D - \zeta N + \lambda_f v_f C_f E_1(W) + \lambda_S v_S C_S E_1(W) \\ - v_P B[P(B, N, W) - \gamma] \qquad (8.4)$$

where D = atmospheric deposition

ζN = leaching

C_f = fast carbon pool

λ_f = decomposition rate of the fast carbon pool

v_f = N:C ratio of fast carbon pool

$E_1(W)$ = effect of water availability on decomposition

= $\exp(\phi W^2)$ (as in CENTURY model), where $\phi < 0$.

C_S, λ_S, v_S give the corresponding parameters for the slow carbon pool

v_P = N:C ratio of plants

$v_P = f v_f + (1 - f) v_S$, where f = fraction of litter being delivered to the fast carbon pool

Note that this equation is for an "open" nitrogen cycle; setting $D = \zeta = 0$ would allow analysis for a "closed" nitrogen cycle. Analysis of the closed cycle, however, will in most cases be much more difficult than analysis of the open cycle; most authors use the open-cycle equations in the subsequent chapters.

Similarly, the equations governing the fast and slow carbon pools in undecomposed organic matter are given by

$$C_f = \mu f B - \lambda_f C_f E_1(W) \tag{8.5}$$

$$C_S = \mu(1 - f)B - \lambda_S C_S E_1(W) \tag{8.6}$$

where f = fraction of litter delivered to fast carbon pool
μ = tissue death, and
$E_1(W)$, λ_f, λ_S are as given above.

Note that this approach omits variable N:C ratios within a plant species, though the overall N:C ratio of the soil could vary as the relative sizes of the fast and slow pools varied, and the average N:C ratio of total aboveground biomass could vary as species abundances changed (i.e., species with different v_P were introduced).

These equations for decomposition and nitrogen mineralization may be viewed as a simplification of the CENTURY model (see Bolker, Pacala, and Parton 1998), though they do omit one feature of the CENTURY model. In CENTURY, low values of soil inorganic nitrogen (N) prevent the transfer of dead organic matter from the low N:C pool to the high N:C pool (the "immobilization" shutdown). We could include this shutdown effect within the same formalism of the simple model, but the immobilization shutdown of litter transfer rarely occurs in the CENTURY model, as plant production shuts down first, usually preventing serious declines in the soil N pool. Thus we ignore the potential for immobilization shutdown of organic-matter transfers in this model.

SUMMARY OF THE BASIC MODEL

The model presented above retains most of the important features of modern ecosystem models, with only a few qualitative omissions (given below). The basic model is given by

$$B = BP[B,N,W] - (\gamma + \mu)B$$

$$P[B,N,W] = \left[\frac{r}{K_L + B}\right]\left[\frac{N^{\theta_N}}{N^{\theta_N} + K_N}\right]\left[\frac{W^{\theta_W}}{W^{\theta_W} + K_W}\right]$$

$$W = R - sW - \frac{BP[B,N,W]}{U} - \varepsilon W$$

$$N = D - \varsigma N + \lambda_f v_f C_f E_1(W) + \lambda_S v_S C_S E_1(W) \\ - v_P B[P(B,N,W) - \gamma]$$

$$C_f = \mu f B - \lambda_f C_f E_1(W)$$

$$C_S = \mu(1 - f)B - \lambda_S C_S E_1(W)$$

This model omits (1) "immobilization shutdown" of dead organic matter transfers, (2) within-species variation in plant N:C ratios, and (3) decreased soil evaporation due to shading.

Guidance was also offered to authors on parameter values within the model, though parameters should vary depending on whether authors are modeling, for instance, forests or grasslands, or mesic or xeric habitats. The parameter values employed are thus given in subsequent chapters.

CHAPTER NINE

Successional Biodiversity and Ecosystem Functioning

Ann P. Kinzig and Stephen Pacala

INTRODUCTION

Any undergraduate in an introductory ecology course can list and explain the different kinds of succession (if keeping up with the material). The principal distinctions of primary versus secondary succession and competition versus facilitation have not changed substantially in several decades, providing some evidence that the broad outline of succession is well understood. Here, we focus on the most widespread and common form—secondary succession driven by competitive interactions among plants.

Consider two different secondary successions in mesic forest on rich soils. In the first, a storm blows down the large trees in a late-successional stand, but spares the more supple or resilient saplings and seeds already present in the site. Resident saplings are most likely to belong to shade-tolerant species, simply because they persisted in the understory before the blow-down. Some early-successional species may be present already, especially if, like pin cherry, they germinate from dormant seed left over from the last disturbance at the site (Marks 1974). Other early-successional species disperse in from surrounding habitat. Because of their rapid growth under the resource-rich conditions following a disturbance, the early-successional species quickly dominate the stand and reduce resource availability, and temporarily suppress slower-growing species. However, the fast-growing species

175

are eventually superseded by the late-successional dominants because the late-successional species can regenerate under lower resource levels or are longer lived.

In the second scenario, an agricultural field is abandoned. Because it entirely lacks a seed and sapling bank of shade-tolerant forest trees, there is an inevitable lag until the late-successional species colonize the site. Species that arrive first become the early-successional dominants, but only because stronger competitors are absent, and they are eventually replaced by the stronger competitors when they arrive.

This first scenario describes what Pacala and Rees (1998) have called the successional niche, while the second describes the competition-colonization trade-off (Levins and Culver 1971; Horn and MacArthur 1972; Armstrong 1976; Hastings 1980; Shmida and Ellner 1984; Crawley and May 1987; Nee and May 1992; Tilman 1994). The critical distinction between the two models is that the late-successional species are the strongest competitors under all conditions under the competition-colonization hypothesis (they always displace the early-successional species when both are present at the same site), whereas they perform less well than the early-successional species in resource-rich conditions under the successional niche hypothesis. The relative importance of the two mechanisms in maintaining successional diversity depends in part on the scale and severity of disturbance. If disturbance spares seeds or juveniles of the late-successional species, or if the disturbed area is small and surrounded by seed sources of late-successional species, then the precolonization lag critical to the competition-colonization mechanism is unlikely to occur. On the other hand, large and severe disturbances, such as stand- and seed-destroying fire or agriculture, will favor the competition-colonization mechanism.

The relative importance of the two mechanisms can also depend on the time scales associated with colonization and growth to maturity. If the fecundities of late-successional species are low enough to permit not only establishment in disturbed patches by early-successional species but also reproduction prior to the arrival of a late-successional species, then the competition-colonization mechanism can operate to permit coexistence. If, on the other hand, the arrival of a late-successional species typically occurs before reproduction by early-successional colonists, then the competition-colonization mechanism is unlikely to occur.

Below we show that ecosystem functioning is typically not maximized under high diversity if coexistence is maintained by the successional niche, and may or may not be maximized under high diversity if coexistence is maintained by the competition-colonization trade-off. Instead, a monoculture often has the greatest performance, with multispecies stands functioning at intermediate levels. The range of ecosystem functioning in two-species stands is typically contained within the range of values from monocultures; values for three-species stands are within the range from two-species stands, and so on.

Why does successional diversity have a different effect on ecosystem functioning than the other forms of coexistence examined in this monograph? In most of the mechanisms that lead to a positive relationship between diversity and ecosystem functioning, competitive ability and ecosystem functioning in a monoculture are positively related. This is true under the sampling hypothesis (chapter 2), where a single species is the best competitor and has the highest performance under all conditions. It is also true under the spatial and temporal niche hypothesis (chapter 10), where species competitively dominate times or local places in which they have the highest function. The niche-partitioning hypothesis differs from the others because it requires that there be no simple competitive hierarchy; species coexist because performance is higher in mixed stands than in monocultures.

We show below that the highest ecosystem functioning is not always associated with the best local competitor under the two successional hypotheses. For example, an oft-repeated scenario for the successional-niche hypothesis is that early-successional species grow fastest under the resource-rich conditions immediately following disturbance, whereas slower-growing late-successional species grow at the lowest resource levels and drive resources to these low levels (as in Tilman 1982). Because of these traits, monocultures of early- and late-successional species may sometimes maximize different ecosystem processes (e.g., productivity for the early-successional species and carbon storage for the late-successional species).

One might think that a function like total carbon storage should be increased by successional diversity because carbon can be stored quickly by early-successional dominants immediately after disturbance and in large amounts by late-successional dominants later on. However, this is not what occurs in the model of successional niches we consider in the following section. In every case, the presence of early-successional species delays dominance of the local site by late-successional species. This effect reduces carbon storage in a successional mosaic of patches more than rapid early growth increases it.

In addition, landscape-level ecosystem functioning is complicated by the number of sites occupied by each successional type, and this often runs counter to expected trends. For example, under the competition-colonization hypothesis, the best competitor for a local site actually has the lowest regional abundance in monoculture, because it has the poorest colonizing ability. The best competitor thus has low landscape-level functioning. But the best competitor also displaces inferior competitors, with higher landscape-level functioning because of their greater monoculture abundance. It is useful to think of a landscape governed by either of the

178

successional niche mechanisms as a mosaic of monocultures, each dominated by a different successional stage. The ecosystem function of the habitat as a whole will thus be a mixture of values for the different monocultures, and so will fall within the range of extreme values for pure monocultures.

THE SUCCESSIONAL NICHE IN A SIMPLE MECHANISTIC ECOSYSTEM MODEL

Consider a mosaic of an infinite number of patches. The community dynamics within each patch are governed by the core ecosystem model of this monograph with one addition. Each species in each patch constantly exports colonizing propagules at rate FB_i, where B_i is the biomass of species-i, and each patch constantly receives the average density of propagules produced in the habitat as a whole. Because fecundity per unit biomass (F) does not vary among species, and because each species has infinite dispersal, the results that follow do not contain any explicit competition-colonization trade-offs. (The competition-colonization trade-off is discussed in the latter half of this chapter.)

In addition, disturbances occur on the infinite collection of patches. When disturbance strikes a patch, its living biomass is transferred to the patch's pools of soil organic matter. The patch's live biomass is then set to zero, and the site must be recolonized from outside. Although the whole collection of patches is physically homogeneous in the sense that the same model parameters apply in each local area, disturbances create heterogeneity in the state variables.

We seek an expression for ecosystem averages across all patches that does not require us to do stochastic simulations for a finite number of patches. Normally, this would be a tall order because the model represents a nonlinear and spatial stochastic process and such processes generally require advanced methods of approximation (Pacala and Levin 1997;

179

Bolker and Pacala 1997, 1999). However, in this case we may derive a highly accurate and simple expression if we omit nitrogen from the model.

First, suppose that after a disturbance, soil water quickly loses "memory" of the conditions in the cell prior to the disturbance. That is, in the water model for a patch,

$$\dot{W} = R - (s + \varepsilon)W - \sum_i B_i P_i(\underline{B}, N, W)/U_i. \qquad (9.1)$$

Suppose that biomasses are suddenly set to zero by disturbance and then increase slowly as colonists arrive. We assume the values of the B's before the disturbance do not affect water levels after the disturbance enough to alter significantly the postdisturbance successional trajectories. In practice, a stochastic simulator constructed from the core ecosystem model will have this behavior, because most realistic values of R, s, or ε will be sufficiently large relative to realistic rates of recolonization. This is because soil moisture typically adjusts at time scales of days to weeks, while vegetation responds at times scales of months to years.

Now consider again our infinite collection of coupled local models with disturbance and without nitrogen. In this model, all patches last disturbed at time t^* will share exactly the same values of the B_i's and approximately the same value of W. This is because B's are set to zero at t^*, and W quickly loses its predisturbance memory. The system is only stochastic because of disturbance and there is no variation among the B's and W's from different patches last disturbed at the same time. We may thus write a pair of deterministic differential equations for $B_i(t^*)$ and $W(t^*)$—respectively, the species-i biomass and soil moisture at time t in patches last disturbed at time t^*—because these quantities behave deterministically:

$$\dot{B}_i(t^*) = B_i(t^*)P_i(\underline{B(t^*)}, \; W(t^*)) - (\gamma + \mu)B_i(t^*) - FB_i(t^*) + \overline{FB_i}$$

$$\dot{W}(t^*) = R - (s + \varepsilon)W(t^*) - \sum_i B_i(t^*)P_i(\underline{B(t^*)}, \; W(t^*))/U_i$$

(9.2)

where $\overline{FB_i}$ is the average seed rain of species-i at time t (full definition below).

Unlike the $B(t^*)$'s and $W(t^*)$, the amount of below-ground carbon will vary significantly among patches last disturbed at the same time. For example, a patch disturbed several times in the recent past and most recently at t^* will have less belowground carbon than a patch that remained undisturbed for many years prior to t^*. However, because the equations for belowground carbon are linear in C_S and C_f, it is correct simply to evaluate them at the average values for patches last disturbed at time t^*:

$$\dot{C}_f(t^*) = \mu \sum_i fB_i(t^*) - \lambda_f C_f(t^*)$$

$$\dot{C}_S(t^*) = \mu \sum_i (1 - f) B_i(t^*) - \lambda_S C_S(t^*)$$

(9.3)

where $C_S(t^*)$ and $C_f(t^*)$ are the average values. Note that if we were to include nitrogen in the model, then the variation in belowground carbon among patches last disturbed at t^* would produce corresponding and long-lasting variation in nitrogen mineralization, and this would induce variation among the B's and W's in those patches. It might work to approximate the nitrogen equation at the average values (see Moorcroft, Hurtt, and Pacala 2000), but it also might not. We thus adopt the conservative approach of deleting nitrogen and its effects altogether.

To complete the model, we still require a way to keep track of how much of the landscape was last disturbed at time t^*, an expression for the average seed rain in equation (9.2) and boundary conditions. Let $Q(t^*)$ be the density at time t of patches last disturbed at time t^* and $D(t^*)$ be the disturbance rate for those patches. Then

$$\dot{Q}(t^*) = -D(t^*)Q(t^*) \tag{9.4}$$

with boundary condition

$$Q(t) = \int_0^t D(t^*)Q(t^*)\,dt^*. \tag{9.5}$$

Boundary condition (9.5) simply says that the density of newly disturbed patches (those for which $t^* = t$) is given as the total that are disturbed at time t. Because dispersal is infinite, the average seed rain for species at time t is

$$\overline{FB_i} = \int_0^{t^*} FB_i(t^*)Q(t^*)\,dt^*.$$

Finally, note that the boundary conditions giving the values of the state variables in newly disturbed patches are

$$B_i(t) = 0$$
$$W(t) = R / (s + \varepsilon)$$
$$C_s(t) = \int_0^t D(t^*)Q(t^*) \tag{9.6}$$
$$[C_s(t^*) + (1 - f)\sum_i B_i(t^*)]\,dt^* \Big/ \int_0^t D(t^*)Q(t^*)\,dt^*$$

$$C_f(t) = \int_0^t D(t^*)Q(t^*)$$
$$[C_f(t^*) + f\sum_i B_i(t^*)]\,dt^* \Big/ \int_0^t D(t^*)Q(t^*)\,dt^*.$$

182

The first condition states that there is no living species-i biomass on newly disturbed land; the second formalizes the idea that soil moisture has faster dynamics than biomass, and hence is unaffected by biomass before the disturbance; and the third and fourth state that the average belowground carbon pools on recently disturbed land are spatially averaged values from the disturbed patches.

It is important to understand that in solving the system (9.2)–(9.6), one obtains precisely the answer that one would obtain from a stochastic simulation of a large number of patches. Mathematically oriented readers should also be aware that this system can be expressed as an equivalent set of VonFoerster (1959) first-order partial differential equations in which the state variables flow forward both in time and in patch age (the time since last disturbance). The system (9.2)–(9.6) describes the *characteristics* of this set of partial differential equations.

Case Studies

In what follows, we restrict attention to the special case in which, because of the relatively fast dynamics of soil moisture, $W(t^*)$ is always at equilibrium with respect to the $B(t^*)$'s. Thus, $W(t^*)$ is the positive root of

$$W(t^*) = \frac{R - \sum_i B_i(t^*) P_i(\underline{B(t^*)}, \ W(t^*)) / U_i}{s + \varepsilon}. \tag{9.7}$$

In addition, we study three separate special cases, each of which is tailored to resemble a different type of ecosystem, within the moderately severe constraints imposed by the structure of the basic model. We illustrate the results using an extremely simple model of disturbance. Patches are disturbed after a fixed time interval, and are uniformly distributed over this interval.

CASE I: MESIC PASTURE/SAVANNA

In Case I, we assume that water is not limiting ($W >> k_w$), which reduces $P_i(\underline{B(t^*)}, W(t^*))$ to

$$P_i(\underline{B(t^*)}, W(t^*)) \approx \frac{r_i}{K_{L,i} + \sum_i B_j(t^*)} . \qquad (9.8)$$

We also assume a trade-off relating a species' maximum rate of local biomass growth (i.e., $r_i/K_{L,i} - \gamma_i - \mu_i$) and its ability to maintain locally positive biomass growth under resource deprivation. Note that a species-i monoculture that happens to escape disturbance for a long time (as t^* becomes large) will approach the steady state: $r_i/(\gamma_i + \mu_i) - K_{L,i}$. If we interpret (9.8) as the response of growth to light availability, then a species' large-t^* steady state in monoculture is related to its R^* (Tilman 1982) for light. As the large-t^* steady state increases, R^* decreases. Thus, the trade-off is between species with large growth rate and R^* and species with the reverse. The former will be early-successional species and the latter will be late-successional species. Obviously, this idea underpins many published works on the ecophysiology of succession (e.g., Bazzaz 1979). Although we examine a number of parameter values, we illustrate results for parameters that cause the model to perform like a successional mix of grassland and invasive shrubs or small trees. The parameters used for the figures below are for a three-species community with: $r_i = 4,2,1$; $\mu_i = 1.80, 0.65, 0.13$; $\gamma_i = 0.25, 0.20, 0.15$; $K_L = 1$; $F = 0.01$; $f = 0.5$; $\lambda_s = 0.04$; $\lambda_f = 1.0$; and disturbances every 15 years.

CASE II: MESIC FOREST

Here, we choose parameters for the model in Case I to yield ecosystem performance measures typical of mesic forest. The parameter values for three types of trees in the examples below were: $\mu_i = 0.33, 0.13, 0.02$; $r = 2$; $\gamma_i = 0.11$,

184

0.05, 0.02; $K_{L,i}$ = 0.4, 0.8, 2.0; F = 0.01; f = 0.5; λ_s = 0.04; λ_f = 1.0; and disturbances every 100 years.

CASE III: XERIC GRASSLAND/STEPPE

In Case III, we assume that low precipitation keeps biomasses well beneath K_L, and keeps soil moisture values well beneath K_W. Thus, (9.8) becomes approximately

$$P_i(\underline{B(t^*)}, W(t^*)) \approx \frac{r_i W(t^*)}{K_{L,i} K_{W,i}}. \tag{9.9}$$

Combining (9.7) and (9.9), we have

$$P_i(\underline{B(t^*)}, W(t^*)) = \frac{r_i R}{K_{L,i} K_{W,i} \left[s + \varepsilon + \sum_j B_j(t^*) \dfrac{r_j}{K_{L,j} K_{W,j} U_j} \right]}. \tag{9.10}$$

We also assume the trade-offs in Case I, but add to them the assumption that rapid maximum net production is inversely correlated with water use efficiency. In this case, we illustrate results for parameters consistent with dry steppe. The parameter values in the examples are again for a three-species community with: r_i = 3, 1, 0.33; μ_i = 0.76, 0.20, 0.03; U_i = 0.0015, 0.001, 0.0005; γ = 0.02; F = 0.01; K_L = 2; R = 400; s = 1.5; ε = 0.365; K_W = 200; f = 0.5; λ_s = 0.04; λ_f = 1.0; and a disturbance interval of 100 years.

Results

The first finding is that the trade-offs examined in this study cannot maintain large numbers of species, at least with the model as defined by equations (9.2) to (9.6). Although we have not been able to prove it, there appears to be a fundamental limit to the similarity of the species that can coexist in the model. This limit is different from the classical limiting similarity of the Lotka-Volterra competition model (MacArthur and Levins 1967). As pointed out by

Abrams (1976), the classical limiting similarity prohibits some combinations of species from coexisting. But the classical models generally permit one to invent specific groups of species that will coexist with arbitrarily close packing. Although we have always thought this distinction moot in practical terms, here it does limit the diversity of examples that we have been able to study.

The closest we have yet come to a proof of a fundamental limit to species-packing in the model is a sufficient (but not necessary) condition for two-species coexistence that holds if the two species are dissimilar enough. Our current numerical record is five coexisting types. Thus, it is probably best to view the results as pertaining to successional guilds, with the understanding that another mechanism must be added to achieve realistic levels of diversity within guilds (such as spatial or temporal heterogeneity in the environment).

Figure 9.1 shows the steady-state pattern of coexistence for Case II. Note that three guilds are present. The long-lived and high-wood-density late-successional dominant increases throughout succession, while the early-successional dominant peaks at approximately 30 years after disturbance. The shortest-lived species exists as a minor component (more like a shrub) that reaches its maximum biomass at 7 to 10 years.

The diversity–functioning relationships predicted for Cases I–III all show a triangular pattern—that is, community-average NPP, total carbon storage, evapotranspiration, and living carbon for Cases I–III (figures 9.2–9.5) are highest and lowest in monoculture, with values for two-species communities contained within the monoculture extremes, and values for three-species communities within the two-species range. The reason for the absence of a clear positive relationship between functioning and diversity is that the successional niche does not involve a consistent trade-off between ecosystem functioning and competitive ability. Early-

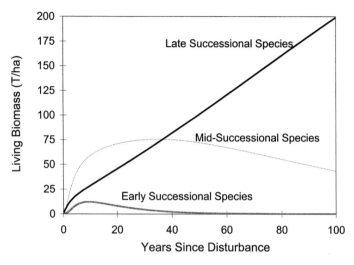

FIGURE 9.1. Steady-state relationship between living biomass and years since disturbance for a landscape containing all three species in Case II.

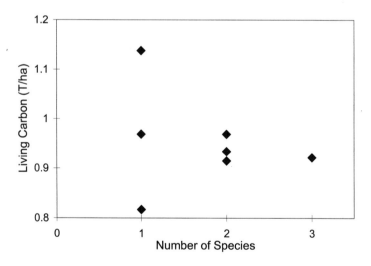

FIGURE 9.2. Steady-state T/ha living carbon in Case I.

187

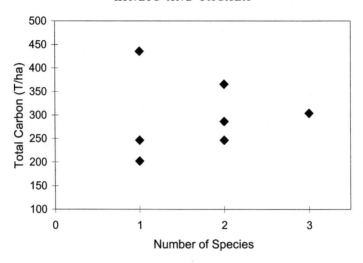

FIGURE 9.3. Steady-state kg/m^2 total carbon (living plus organic matter) in Case II.

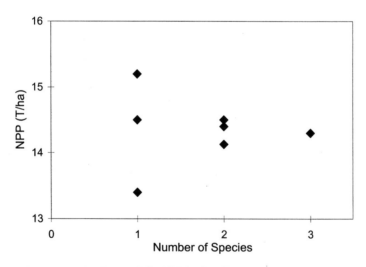

FIGURE 9.4. Steady-state T/ha NPP in Case II.

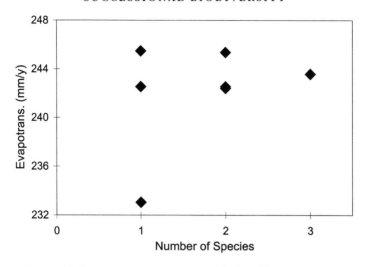

FIGURE 9.5. Steady-state evapotranspiration in Case III.

successional species dominate initially at the expense of the remaining species because they maximize ecosystem functions associated with rapid growth under the resource-rich conditions typical of recently disturbed sites. Late-successional species dominate late because they maximize functions associated with persistence under resource-poor conditions. If diversity is maintained solely by a successional niche, then the average functioning of a successional landscape is the average of the functions of the successional stages present across the landscape. Patches dominated by early-successional species with high NPP, low water-use efficiency, and low carbon storage are averaged with patches dominated by late-successional species with the reverse, producing intermediate average functioning.

This same explanation also applies to the response of the model ecosystem to a sudden environmental change (figure 9.6). A step increase in the r's (perhaps due to fertilization), increase in the μ's (say increased harvesting), or decrease in rainfall always produced the greatest range of change in

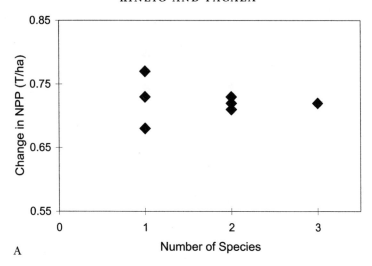

A

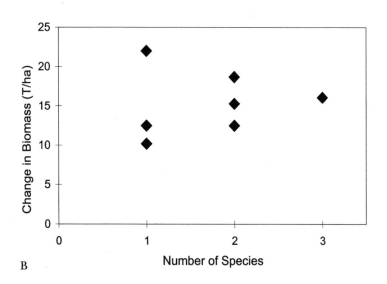

B

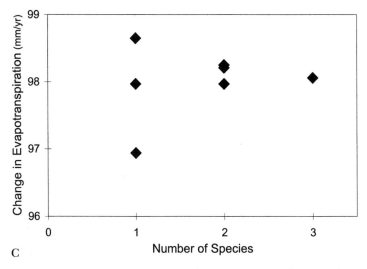

C

FIGURE 9.6. Effects of diversity on the response to environmental change. (A) Change in steady-state NPP (T/ha) resulting from a 5% increase in the growth rate r in Case II. (B) Change in living biomass (T/ha) resulting from a 5% increase in r in Case II. (C) Decrease in evapotranspiration (mm/y) resulting from a 100 mm/y decrease in precipitation in Case III.

function among the three monocultures, with two-species responses contained within the range produced by the monocultures, and the three-species response contained within the range produced by the two-species responses. The average sensitivity of a successional landscape to environmental change is simply the average of the sensitivities from the different successional stages, if diversity is maintained solely by the successional niche.

The result that ecosystem functioning is not maximized by diversity, if diversity is maintained solely by the successional niche, is perhaps not too surprising because succession is a deterministic directional change. Species designed to grow quickly under resource-rich conditions do not store large amounts of carbon under resource-poor conditions. Species that invest in the structural material necessary for

191

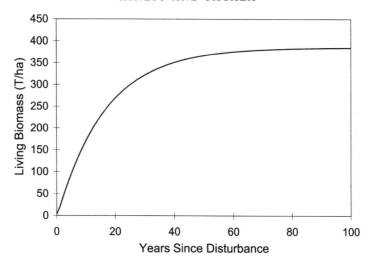

FIGURE 9.7. Steady-state relationship between living biomass and years since disturbance for a landscape containing only the late-successional species in Case II.

long lifespan cannot grow quickly. On the other hand, the results in figures 9.2–9.5 were initially surprising to us, because we expected that successional diversity would maximize carbon storage by combining species capable of rapid growth early in succession with species capable of large carbon storage late in succession. The reason that this does not occur in the model developed here is that the early-successional species delay dominance by the late-successional species. The successional delay reduces carbon storage more than rapid early growth increases it.

To see this, compare figure 9.1 with figure 9.7. The latter figure was produced by running the model for a monoculture of the late-successional species shown in figure 9.1. Note that the intermediate species in figure 9.1 delays dominance by the late-successional species, and that this delay strongly reduces carbon storage during succession. In particular, the site containing only the late-successional species reaches an equilibrium total living biomass of about 375 T/ha after about 75 years (figure 9.7), while the site containing early-, mid-, and late-successional species is only storing about 240 T/ha after 100 years.

COMPETITION-COLONIZATION IN A SIMPLE MECHANISTIC ECOSYSTEM MODEL

Like the successional-niche model, the competition-colonization model assumes a mosaic of an infinite number of patches. Again, the common ecosystem model is utilized, with one addition. Different species are assumed to devote a different proportion (φ_i) of their net primary production to seed growth, and the proportion thus devoted is assumed to be unavailable to fuel further production. Thus, fecundity varies among species as $\varphi_i \text{ NPP}_i$, and the equation for growth of the ith species (in monoculture) in the common ecosystem model becomes

$$dB_i/dt = (1 - \varphi_i)B_i\left[\left(\frac{r}{K_L + B_i}\right)\left(\frac{N}{K_N + N}\right)\left(\frac{W}{K_W + W}\right) - \gamma\right] - \mu B_i. \quad (9.11)$$

This is the only performance trade-off required to maintain a competition-colonization differentiation in the common ecosystem model; species that devote a higher proportion of their NPP to seed production can exploit a greater number of empty sites, but these same species are competitively excluded by those species devoting a lesser fraction of their production to fecundity when both occupy the same site. One might imagine that other performance trade-offs would accompany the competition-colonization trade-off—for instance, the high-fecundity, "weedier" species might have high initial growth rates (r_i) and higher rates of respiration (γ_i) relative to the late-arriving superior competitors. These types of performance trade-offs begin to approximate characteristics employed in the successional niche model dis-

193

cussed above, but will be considered within this section as well.

The features of the competition-colonization model have been thoroughly discussed by Levins and Culver 1971; Horn and MacArthur 1972; Hastings 1980; Nee and May 1992; Tilman 1994; May and Nowak 1994; and Kinzig et al. 1999, among others. This class of models assumes that there is an inverse relationship between competitive ability and colonization ability, and that there is a strict hierarchy in which competitively superior species exclude competitively inferior species when both occupy the same site. Moreover, this competitive exclusion is assumed to be instantaneous. These assumptions translate into the dynamic model

$$\frac{dp_i}{dt} = \chi_i p_i \left(1 - \sum_{j=1}^{N} p_j\right)$$

$$+ \chi_i p_i \sum_{j=i+1}^{N} p_j - p_i \sum_{j=1}^{i-1} \chi_j p_j - m p_i \qquad (9.12)$$

where p_i = fraction of sites occupied by the ith species and m_i = mortality rate of the ith species (taken here to be equal for all species, $m_i = m$). Competitive ability decreases with increasing rank i, and species produce propagules at a rate given by $\chi_i p_i$ ($\chi_i > \chi_j$ for $i > j$); these propagules can fall on empty or occupied sites.

Note that the abundance of a particular species is only affected by itself and its superior competitors; superior competitors are indifferent to the presence of inferior competitors. Furthermore, limiting similarity can emerge from this model—if a superior competitor is present, the fecundity of the next-ranked competitor must differ from the fecundity of the superior competitor by a finite amount, or coexistence cannot occur. The amount by which a species' fecundity must differ depends on the abundances of all superior species present. Once the threshold fecundity for coexistence is achieved, abundance will increase with increasing fecundity (assuming no other superior species invade the system).

In this analysis, we extend the assumption of instantaneous competitive exclusion, and assume that equilibrium values of standing biomass, soil water, soil nitrogen, and carbon are "instantaneously" achieved when a site is colonized. In other words, we assume growth and soil dynamics are fast relative to disturbance and colonization events. This assumption is consistent with the conditions specified in the introduction under which the competition-colonization mechanism might hold. In particular, growth dynamics and water and nitrogen dynamics will likely be rapid relative to disturbance and colonization if the competition-colonization mechanism is operating; this assumption extended to the slow carbon pool is a bit more problematic, but we employ it here for tractability's sake.

Under these assumptions, then, fecundity (χ_i) can be related to proportion of NPP devoted to seed production (φ_i) by

$$\chi_i = Mg\mu \frac{\varphi_i}{(1 - \varphi_i)} \tag{9.13}$$

where NPP at equilibrium is given by $B_i \, \mu/(1 - \varphi_i)$ (eq. 9.11 above), the mass of an individual seed is given by B_i/M, and g is some germination "tax" (not all seeds produced are viable; $0 <= g <= 1$).

Local versus Global Performance

Under the assumption of instantaneous achievement of equilibrium states following colonization, it is straightforward to find monoculture equilibrium values for B_i, $C_{i,f}$, $C_{i,S}$, N, and W, as well as equilibrium values for the ecosystem processes of interest (NPP, total carbon, N-mineralization, and evapotranspiration). We use the common ecosystem

195

model as presented in chapter 8 (with the exception that $E_1(W)$ is set to 1) with the modification to the growth equation as given in (9.11). These equations give site-specific equilibrium values; average landscape-level values for stocks and processes are found by multiplying the site-specific equilibrium values for the various types by their landscape abundances given by equation (9.12). Note, however, that for the nitrogen equation of the common ecosystem model to hold, there must be some source of incoming nitrogen that exactly balances the nitrogen exported from a site in seeds. We impose this restriction here in order to allow comparison with the closed nitrogen cycle used in the previous section and in other chapters (i.e., N is still determined by the balance between leaching and deposition), but this assumption is relaxed in some of the cases presented below.

These ecosystem processes—at a particular site—are graphed as functions of φ_i in figure 9.8. Note that the superior competitors (low allocation to seed production) are also the superior performers for total carbon, N-mineralization, and evapotranspiration, while the inferior competitors (good colonizers) have the highest net primary production across the range of fecundities considered here. Thus, the superior competitor is not always the superior performer at a particular site in this system.

This picture changes, however, when one considers landscape-level function in a monoculture. By definition, the superior competitor is constrained in its colonization ability; it *always* (in this model) occupies fewer sites in monoculture than would an inferior competitor in monoculture. Thus, when one averages function across the landscape and across empty and occupied sites, both NPP and evapotranspiration are maximized for inferior competitors, while total carbon and N-mineralization are maximized for intermediate competitors (figure 9.8b). At the landscape level in monoculture, the superior competitor is *never* the superior performer for the processes of interest here. (This pattern may change for polycultures, since the abundance of inferior competitors will be reduced by the presence of superior competitors. Thus, in a multispecies community, the superior competitor may, under some circumstances, make the largest contribution to certain ecosystem processes, and/or may be present in the best-performing community.)

Cases Considered

We consider three cases below, spanning a range of possible community structures, additional (beyond competition-colonization) performance trade-offs, and closed versus open nitrogen cycles. In each case, a 10-species pool is used to examine the relationship between diversity and ecosystem process, and landscape-level values of processes are calculated for all possible 1-, 2-, 3-, 8-, 9-, and 10-species combinations drawn from the pool. The complete species pool is chosen such that all 10 species can coexist, but note that this does not guarantee that all possible n-species combinations can coexist ($2 \leq n \leq 9$). (Recall that the ability of an inferior competitor to establish itself in a certain community depends on the abundances of superior competitors. A shifting complement of superior competitors under different levels of diversity can create conditions under which inferior competitors cannot persist in the system.) Note that this situation violates one of the central assumptions of the sampling hypothesis—namely, that the superior performer is more likely to be present in a high-diversity community than in a low-diversity one precisely because all n-species combinations are assumed possible. (On the other hand, if drawn to be part of the n-species community in the model system, the superior competitor always survives—thus obeying at least this assumption of the sampling-effect model.)

In empty sites, evapotranspiration is taken to be equal to evaporation, the soil carbon pools are calculated by using the average time to colonization and assuming decomposition has occurred over this time, and N-mineralization rates

197

are calculated using these carbon pools. NPP is taken to be 0 on empty sites. Equilibrium values of ecosystem processes on occupied sites are calculated using the equations above, and landscape-level processes (on a per-area basis) are calculated by averaging over all empty and occupied sites.

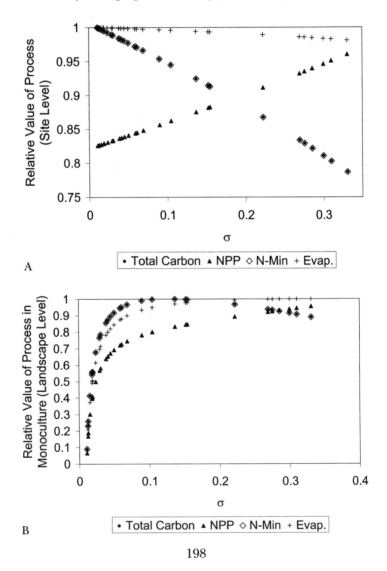

A

B

CASE 1: COMMUNITY/STRUCTURE (CS)

There are many different ways to construct 10-species communities in which the fecundities meet the constraints imposed by the hierarchical competition-colonization model. Do species invade with the lowest possible fecundities that can be maintained in the system (in which case they have very low abundances)? Is some minimum viable abundance required? What are the broader regional processes constraining the pool of species available to colonize? In order to test the effects of community structure, we constructed three different 10-species communities. In the first, (E, even) fecundities are chosen so that in the 10-species community, each species has an abundance of about 8%. In the second (R, random), fecundities are chosen at random between a predetermined minimum and maximum rate (with the random choices continuing until a coexisting 10-species pool is achieved). In the third (S, structured), fecundities are chosen so that there is a "structure" to the abundances in the 10-species community—the superior competitor has an abundance of 20%, the next has an abundance of 16% (20% of 80%), and so on. Parameters for the common ecosystem model are as given in figure 9.8a; the only performance trade-off is competition-colonization (spe-

---

FIGURE 9.8. (A) Ecosystem processes versus proportion of net primary production devoted to seed production (φ) at a particular site in the competition-colonization model. Note that total carbon and N-mineralization are proportional to one another. Parameters in the common ecosystem model are as follows: $s = 1$, $\mu = 0.1$/year, $\varepsilon = 0.2$, $f = 0.25$, $\gamma = 0.1$/year, $K_L = 1$ kg/m^2, $K_N = 0.005$, $K_W = 100$ mm/year, $\lambda_f = 1$, $\lambda_s = 0.04$, N^* ($= D/\xi$) $= 0.0066$, $r = 3$ kg/m^2, $R = 750$ mm/year, $U = 0.0015$ kg/(m^2 mm), $v_f = 0.0666$, $v_s = 0.00444$. Parameters for relating fecundity to φ are: $M = 1000$, $g = 0.01$. In the competition-colonization equation, m is taken to be 0.01. (B) Ecosystem processes versus proportion of net primary production devoted to seed production across the landscape in the competition-colonization model. Note that total carbon and N-mineralization are proportional to one another. Parameters are as given in part A.

199

cies differ only in φ_i). Fecundities, monoculture abundances, and abundances in the 10-species community are given in table 9.1.

CASE 2: PERFORMANCE TRADE-OFFS (PF)

In addition to species having differential competition-colonization abilities, we assume in this analysis that species differ in other performance characteristics as well. In particular, we consider variations in r (Eq. (9.11), growth rates increase with decreasing competitive ability), K_N (K_N increases with decreasing competitive ability), v_P (N:C ratio of the plant, affecting allocation to fast and slow carbon pools upon plant death, superior competitors are "woodier" than inferior competitors), and γ (respiration increases with decreasing competitive ability). In this analysis, we consider only the R community above (randomly generated fecundities); the superior competitor in that system has a fecundity of 0.012 (against a disturbance rate of 0.01) and the inferior competitor (10th ranked) has a fecundity of 0.408. The corresponding ranges for the performance parameters are: r (2.5 to 3.75), K_N (0.003 to 0.009), v_P (0.076 to 0.05), and γ (0.08 to 0.256). Parameter values for all intermediate competitors were scaled according to relative fecundities.

CASE 3: CLOSED NITROGEN CYCLE (T)

All parameters are as in Case 2 above, except that D (deposition) and ξ (leaching) have been set to 0. In addition, an amount of nitrogen equal to that exported in seeds is delivered to the organic soil pools, so that the total nitrogen (T) at a site remains constant. Soil nitrogen levels are now controlled by mineralization and uptake, rather than deposition and leaching, and thus can vary depending on community composition. Note that we assume, for simplicity's sake, that the exported seeds have the same C:N ratio as the rest of the plant, and that incoming seeds are allocated to both fast and slow carbon pools. We have also, for tract-

Table 9.1. Fecundities Used in Case 1 of the Competition-
Colonization Analysis for Different Community Structures, and
Associated Abundances

Rank	Fecundity (χ_i)	Monoculture Abundance (%)	Abundance in 10-species Community (%)
Community E: Even Abundance			
1	0.0109	7.8	7.8
2	0.0129	22	8.1
3	0.0156	36	7.9
4	0.0193	48	8.2
5	0.0245	59	7.9
6	0.0320	69	8.0
7	0.0435	77	7.9
8	0.0630	84	8.2
9	0.0990	90	7.8
10	0.1800	94	8.3
Community R: Random Fecundities			
1	0.0120	17	17
2	0.0178	44	16
3	0.0236	58	5.2
4	0.0274	64	4.2
5	0.0347	71	9.0
6	0.0470	79	6.3
7	0.0628	84	6.2
8	0.1461	93	18
9	0.3308	97	2.2
10	0.4082	98	1.3
Community S: Structured Abundance			
1	0.0125	20	20
2	0.0195	49	16
3	0.0304	67	13
4	0.0475	79	10
5	0.0745	87	8.3
6	0.1170	91	6.6
7	0.1830	95	5.2
8	0.2860	97	4.2
9	0.4500	98	3.4
10	0.7100	99	2.7

ability's sake, removed water limitations in this system, and thus do not consider diversity-evapotranspiration results for this particular case.

Results

CASE 1: VARIATIONS IN COMMUNITY STRUCTURE

The results for Case 1 (communities E, R, and S) are shown in figures 9.9–9.11. Most of the relationships appear to be of the "general" form that would be expected under the sampling or niche complementarity hypotheses (functioning initially increasing with increasing diversity, and then saturating) when one looks at the average values of function within the set of all n-species communities (large circles in the figures). In the S system, however, there appears to be little or no relationship between diversity and NPP (NPP independent of diversity, on average), and in the R case, both NPP and evapotranspiration show a slight tendency toward an inverted U relationship of diversity with function—that is, there appears to be a slight downturn in function at higher levels of diversity relative to intermediate levels of diversity.

In terms of the "envelope" of points (i.e., the trends for the best- and worst-performing communities as diversity increases), the sampling hypothesis would predict a flat top—that is, the best possible performance will not depend on diversity, while the niche complementarity model would predict an increasing upper bound (see Tilman and Lehman, chapter 2; Hector, chapter 4). In almost all of the cases presented here, however, we see a decline in the best performance with increasing diversity (though in many of the cases this decline is slight). The best-performing 10-species community nearly always has a lower performance than the best-performing monoculture, and we see a particularly clear decline in best performance with increasing diversity in the case of NPP. Moreover, the members of the best-performing community are not static. Consider, for instance, total carbon in the S community. The best-performing monoculture consists of species of rank 6, the best two-species community consists of species 4 and 10, the best three-species community consists of species of rank 3, 6, and 10, and so on. The expected results under the sampling and niche complementarity hypotheses, in contrast, are that the species with the best performance in monoculture will continue to be present in the best-performing higher-diversity communities.

To understand these results, we must return to the differences between performance and competitive ability at both a site and a landscape level. Recall that the best monoculture performers were not the superior competitors (figure 9.8b). In polycultures, however, the abundance—and thus the landscape-level performance—of those species that do best in monoculture may be suppressed by the presence of superior performers. Thus, those species present in the best-performing monocultures may fail to contribute to the best-performing communities in polyculture. Moreover, a two-species community may actually have lower total occupancy of sites than many monocultures, further suppressing landscape-level functioning.

Thus, many of the assumptions present in the sampling and niche complementarity hypotheses—a single type consistently contributes to the best-performing community in spite of changes in number of species within the community, and all possible n-species communities drawn from the pool of N species can exist—are violated in this system. In spite of these violations, we see average diversity–function relationships that are similar to those predicted under these hypotheses. The envelopes of function versus diversity, however, frequently differ from those expected under the prevailing hypotheses, both in shape and in the identity of the species contributing to the top-performing community.

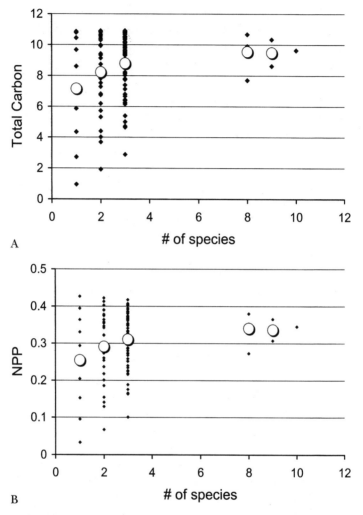

FIGURE 9.9. Ecosystem processes versus diversity for the "Even" community (fecundities chosen so that all types have abundance of about 8% in a 10-species community). Diamonds indicate outcomes for a particular *n*-species configuration, while open circles give the average functioning within a diversity level. Parameters for the common ecosystem model are as given in the caption to figure 9.8, and fecundities are given in table 9.1. Units are as follows: kg/m^2 (carbon storage), kg/m^2/year (NPP), kg(N)/m^2 (N-mineralization), and mm/year (evapotranspiration).

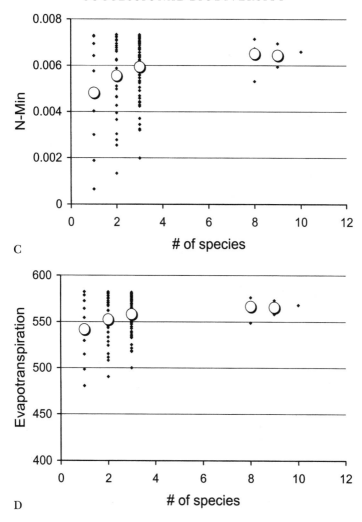

C

D

CASE 2: OTHER PERFORMANCE TRADE-OFFS

The results for Cases 2 and 3, relative to those obtained for Case 1, are shown in table 9.2. In the case of additional performance trade-offs (beyond differences in allocation of NPP to seed production), the behavior of the upper-bound performance with diversity changes for all four functions.

205

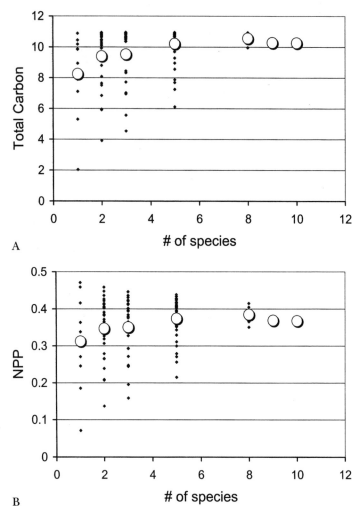

A

B

FIGURE 9.10. Ecosystem processes versus diversity for the "Random" community (fecundities chosen at random within a certain range until a 10-species coexisting community is achieved). Diamonds indicate outcomes for a particular n-species configuration, while open circles give the average functioning within a diversity level. Parameters for the common ecosystem model are as given in the caption to figure 9.8, and fecundities are given in table 9.1. Units are as given in figure 9.9.

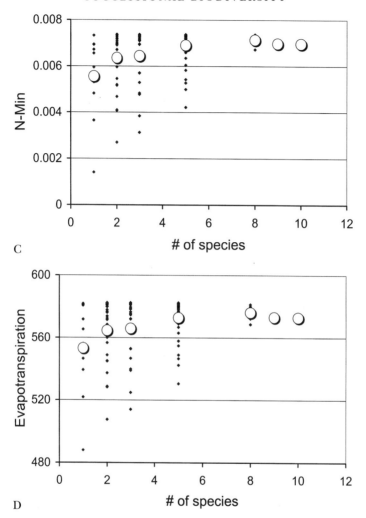

C

D

These results can be related to changes in site-level and landscape-level performances as a function of competitive status. In particular, at the site level, total carbon, NPP, and evapotranspiration now all decline with declining competitive ability (the superior competitor is also the superior site-level performer), while N-mineralization increases with de-

207

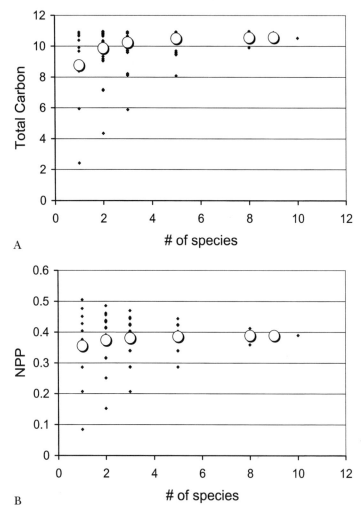

FIGURE 9.11. Ecosystem processes versus diversity for the "Structured" community (fecundities chosen so that the superior competitor has the highest abundance in a 10-species community, the next-best competitor has the next highest abundance, etc.). Diamonds indicate outcomes for a particular n-species configuration, while open circles give the average functioning within a diversity level. Parameters for the common ecosystem model are as given in figure 9.8, and fecundities are given in table 9.1. Units are as given in figure 9.9.

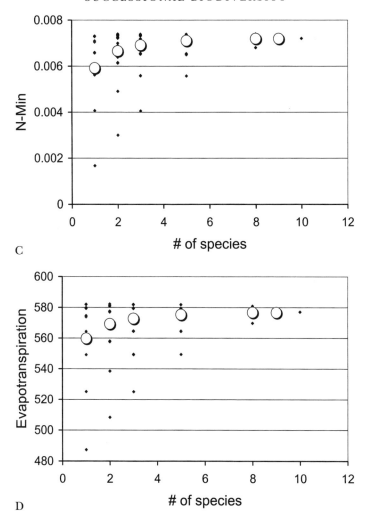

C

D

clining competitive ability (contrast with the previous case, figures 9.9–9.11). In addition, performance differences are now greater across the range of fecundities considered. Furthermore, at the landscape level, total carbon, NPP, and evapotranspiration now peak with an intermediate competitor (rank 6, 7, and 8, respectively, in this case), while

209

TABLE 9.2. Forms for Diversity-Functioning Relationships, and Upper Bounds of Performance, under Various Assumptions

Function	Feature	Case R	Case 2: Other Performance Trade-offs	Case 3: Closed Nitrogen Cycle
Total C	Average performance	Increasing[a,b]	Increasing	Increasing
	Envelope (upper bound)	Flat[c]	**Increasing (weak)[e]**	Increasing
NPP	Average performance	Increasing (weak)[b,d]	Increasing	Increasing
	Envelope (upper bound)	Declining	**Increasing (weak)[e]**	Increasing
N-min	Average performance	Increasing[b]	**No dependence on diversity**	No dependence on diversity
	Envelope (upper bound)	Flat[c]	**Declining**	Declining
Evap	Average performance	Increasing (weak)[b,d]	Increasing (weak)[d]	NA[f]
	Envelope (upper bound)	Flat[c]	**Increasing (weak)[e]**	NA

Notes: Boldface indicates a qualitative change in either the average performance or the envelope under a change in assumptions.

[a] "Increasing" relationships are those where functioning (when averaged across all communities within a diversity level) initially increases with increasing diversity, and then levels off.

[b] In these cases, there is actually a slight downturn in function at higher levels of diversity—that is, the average performance for the 9- and 10-species communities are actually lower than the average performance for the 8-species community. At lower levels of diversity, there is an increase of ecosystem process with diversity.

[c] The best performance is relatively flat over the range of 1- to 8-species communities, but declines with 9- and 10-species communities.

[d] The increase in process is labeled "weak" if the average performance of all monocultures is within 15% of the average performance of all 10-species communities.

[e] The decline or increase in best performance is labeled "weak" if the best performance of the 10-species community and monoculture are within 15% of each other.

[f] The water model was excluded in this analysis, and thus diversity-process relationships for evapotranspiration were not considered.

N-mineralization continues to increase with decreasing competitive ability (maximum levels of performance in monoculture achieved with species of rank 10) (again, contrast with figure 9.8b).

In the case of total carbon, NPP, and evapotranspiration, then, the performance differences between superior and intermediate-ranked competitors are now great enough that adding these superior competitors to the best monoculture improves performance, and so on through increasing levels of diversity. In addition, adding inferior competitors to the best monoculture can also improve performance, as the performance of the intermediate-ranked species is not affected by the inferior competitor (though coexistence in such polycultures is not always possible—see discussion above). For N-mineralization, however, superior competitors are inferior performers at a particular site, so adding superior competitors to the best monoculture depresses the performance of the intermediate competitor without compensating for this loss. Thus, best performance declines with increasing diversity. This decline for N-mineralization is enough to affect average performance, which no longer shows a relationship to diversity.

CASE 3: CLOSED NITROGEN CYCLE

Closing the nitrogen cycle does not lead to any additional differences in the diversity-process relationships beyond those manifested in Case 2 (see table 9.2). The same qualitative patterns of site-level and landscape-level functions across fecundities hold for Cases 2 and 3. The increases in best performance with diversity are no longer "weak" in this case, however, as closing the nitrogen cycle (and removing water limitations) increases the relative difference in site-level performance between superior and inferior competitors.

CONCLUSIONS

Unlike the findings in the theory to date (see review in chapter 2), or the results in the theory chapters to follow (chapters 10 and 11), the successional niche and competition-colonization analyses point to the possibility of declining functioning with increasing species diversity. This can be attributed to the relationship between functional performance and competitive ability. In the competition-colonization model, the superior competitor does not always exhibit the highest functioning for the ecosystem processes considered here, particularly at the landscape level. The presence of the superior competitor, moreover, will depress the higher functioning of the intermediate competitors by reducing their abundance, and thus reducing their contributions to landscape-level processes. Similarly, in the successional niche model, the early-successional species delay the growth of the better-performing, late-successional species; including early-successional species in a community mix actually depresses the functioning of the mixed stand relative to a monoculture containing only the late-successional species.

These findings are relatively robust for the successional niche models. In the competition-colonization models, on the other hand, any of the three major relationships between diversity and ecosystem functioning can be found— increasing functioning with diversity, decreasing functioning, or no effect—depending on assumptions made about community structure, and other performance trade-offs that might accompany the competition-colonization trade-off. Nonetheless, the results of this chapter indicate that there is an important class of models of multispecies coexistence for which the requirements of coexistence and niche partitioning are such that increasing diversity may lead— under reasonable circumstances—to declines in ecosystem functioning.

CHAPTER TEN

Environmental Niches and Ecosystem Functioning

Peter Chesson, Stephen Pacala,
and Claudia Neuhauser

INTRODUCTION

The physical environment is strikingly variable in time and space, providing challenges and opportunities for the organisms in any ecosystem. At first thought, such temporal and spatial variation might be expected to be disruptive to ecosystem functioning. However, the extent to which this is so must depend on the structure of the ecosystem. At the ecosystem level, we can ask, What properties of the organisms individually and collectively maximize ecosystem functioning in the presence of environmental variability? At the level of individual organisms, we can ask, How does natural selection acting on individuals affect ecosystem functioning in the communities in which the individuals belong? Some ecological processes are naturally related to fitness. High primary productivity, for example, might arise when there is a selective advantage for individual plants to have high productivity. On the other hand, carbon storage, as an ecosystem function, may not be linked to fitness when it involves accumulation of dead or decayed rather than productive biomass. Moreover, other strategies for individual success, such as resistance to mortality agents, may tend to have a negative relationship with productivity.

Due to the presence of fitness trade-offs, a species living in a variable environment is unlikely to perform well under all circumstances even if highly phenotypically plastic or consisting of a wide range of genotypes capable of specializing on particular conditions as they arise. Environmental conditions not efficiently exploited by any given set of species provide opportunities for other species, better adapted to those conditions, to make use of resources available at those times or in those places. If specialization on particular environmental conditions makes an individual species more productive, a high diversity of environmentally specialized species should make the system as a whole more productive. Thus, ecosystem service correlates of high productivity should be maximized also at high diversity. These conclusions, however, require that species be present at high enough densities, or be able to grow rapidly enough, to make full use of the available opportunities as they arise.

Underexploited environmental niches, that is, environmental conditions for which the species present are not making maximum use of the available resources, ought to fill with time through invasion of new species or adaptive radiation. Given enough time, with a frequency of variation in environmental conditions that is sufficiently predictable, at least on a long timescale, ecosystems might be expected to fill with species specialized to a range of particular environmental conditions, provided these different species are able to coexist with one another. Such filling of environmental niches should also have the effect of minimizing temporal variation in ecosystem functioning.

As the physical environment varies temporally on many timescales, an important question is how well a system that is shaped on a particular timescale continues to function when exposed to longer-term change. The organisms in a particular ecosystem may be adapted to year-to-year fluctuations in weather, and the system may be saturated with species that coexist in the presence of such year-to-year variation. When the physical environment changes over the longer term—so that average environmental conditions, or the frequency of extreme conditions, are changed—will ecosystem functions, such as primary production, dramatically decline? If the species present in a system are adapted to a range of environmental conditions, then in the face of change, species adapted to the new conditions may be present, and may continue to support ecosystem-level functioning (Tilman and Downing 1994). Species' relative abundances are likely to shift, favoring the species better adapted to the new conditions, and there are likely to be fewer species or a less even distribution of species adapted to the new conditions. In effect, the diversity of the system would be lowered. Invasion from other habitats, and longer-term evolutionary changes within species, may allow restoration of that diversity.

These various ideas are explored here in models of environmental niches. We first ask what environmental niches are, and in what manner they allow the coexistence of a diverse community of species adapted to a range of environmental conditions. We then go on to consider how ecosystem functioning within such systems varies with the diversity of the species pool, and the diversity of a community assembled from the species pool. Finally, we consider how ecosystem functioning of these systems responds to long-term environmental change.

ENVIRONMENTAL NICHES

Species adapted to different environmental conditions may be thought of as having different environmental niches. Environmental niches have both temporal and spatial aspects corresponding to the temporal and spatial aspects of environmental variation. Using an analysis of variance approach (Chesson 1985), we can think of environmental variation as divisible into a purely temporal component (the main effect of time), a purely spatial component (the main effect of space), and a purely spatio-temporal component (the space-by-time interaction). Environmental niches may therefore be classified in accordance with the kinds of environmental variation that they involve. Thus, niches might be temporal, spatial, spatio-temporal, or some mixture of these three types, with consequences for species coexistence and also, as we shall see below, for stability of ecosystem functioning in the face of environmental change.

Temporal Niches

The temporal aspect of an organism's environmental niche is defined by the organism's temporal pattern of activity; for example, when and how intensely the organism is photosynthesizing, germinating, growing, foraging, reproducing, or remaining dormant. That pattern of activity may be defined on short timescales, such as a day, or on longer timescales, such as a year, but it extends to decades and centuries for long-lived organisms like trees. Patterns of activity may be cued by strictly periodic phenomena such as photoperiod or less predictable but still periodic phenomena such as temperature, or they may represent physiological responses to environmental conditions or resources, such as changes in individual growth rates due to changes in temperature or variation in rainfall, or simply development time independent of the external environment, as seen most strikingly in periodical cicadas. The full temporal niche also includes temporal variation in mortality rates even though mortality is not so naturally thought of as "activity." Temporal activity patterns vary from strongly deterministic, for example, cueing by photoperiod (Rathcke and Lacey 1985), to highly stochastic, for example, cueing of germination by specific weather conditions occurring unpredictably from year to year in arid environments (Juhren, Went, and Phillips 1956; Loria and Noy-Meir 1979–1980; Rathcke and Lacey 1985; Bowers 1987; Baskin, Chesson, and Baskin 1993).

In a community context, theory predicts that separation

216

of temporal niches promotes species coexistence. It says that even species with identical resource requirements may coexist when there are appropriate differences between their temporal patterns of use of resources (Grubb 1977; Chesson and Warner 1981; Abrams 1984; Shmida and Ellner 1984; Brown 1989; Loreau 1992; Chesson 1994). A general study of species coexistence in temporally variable environments identifies two general mechanisms of coexistence associated with temporal patterns of activity (Chesson 1994). This study seriously limits the possibilities of other such mechanisms. These two mechanisms are known as *relative nonlinearity of competition* and *the storage effect*. Relative nonlinearity of competition involves different nonlinear responses of different species to fluctuating resources. Under this mechanism, there need be no direct responses of organisms to variation in the physical environment. The storage effect involves direct responses of the organisms to variation in the physical environment but does not require different species to have different nonlinear responses to resources. In our usage, resources are consumable items, such as water, while the physical environment consists of nonconsumable properties such as temperature. Of the two mechanisms, the storage effect appears by far the stronger and more general mechanism of coexistence (Chesson 1994) and closely corresponds to the intuitive concept of coexistence by separation of temporal niches. However, recent evidence (Huisman and Weissing 1999) suggests that relative nonlinearity of competition may be important when there are several limiting resources that are not substitutable (*sensu* Tilman 1982). Study of the storage effect emphasizes that it is not sufficient for species simply to have different temporal activity patterns (*species-specific responses to the environment*) for coexistence to be promoted. Indeed, the storage effect has been shown to have two other requirements, termed *covariance between environment and competition* and *buffered population growth* (Chesson 1994). These three requirements for the storage effect can be explained with reference to the common ecosystem model (chapter 8).

To modify the common ecosystem model to include temporal niches, the per-unit rate of biomass change for a given species j, can be written

$$\frac{1}{B_j} \cdot \frac{dB_j}{dt} = r_j(t) = E_j(t)P[B., N, W] - \delta \qquad (10.1)$$

where $\delta = (\gamma + \mu)$, $B.$ is the sum of the biomasses of the n species in the system and

$$P[B., N, W] = \frac{a}{B. + K_L} \cdot \frac{N}{N + K_N} \cdot \frac{W}{W + K_W}. \qquad (10.2)$$

Here, r of the common model has been replaced by $E_j(t)a$, with $E_j(t)$ representing temporal variation in growth activity by species j, and a representing a reference value of r—for example, a species and time average of r. The $E_j(t)$ define multiplicative departures of r from a due to species and time. We refer to $E_j(t)$ as the *environmental response of species j*. The pattern of variation in the environmental response, $E_j(t)$, over time, defines the temporal niche of a species. Different species have different patterns of response to temporal variation, which separate their temporal niches and provide the first important requirement for the storage effect, species-specific responses to the environment. These different patterns may simply reflect different responses to seasonal change, which may be represented deterministically as different periodic functions of time, or they may be stochastic, perhaps representing different responses of the species to common variation in the weather.

The function $P[B., N, W]$ summarizes the species responses to resources. If the $E_j(t)$ were all equal to 1, this system would have neutral equilibria at which all species densities remain constant and satisfy the equation $P[B., N,$

$W] = \delta$. Under nonequilibrium conditions, the value of $P[B., N, W]$ would reflect competition between the species, among other things; but in any case, in this neutral situation, we can expect competition to tend to restore the value of $P[B., N, W]$ to δ following perturbation from equilibrium. We can therefore use the difference

$$C(t) = \delta - P[B., N, W] \qquad (10.3)$$

as a measure of the magnitude of competition at any given time relative to the magnitude at equilibrium. The quantity (10.3) is sometimes referred to as the *competitive response* (Chesson and Huntly 1997), but we shall refer to it here simply as *competition*. As defined here, it is a measure of overall resource availability. Its magnitude defines restrictions on growth due to resource shortage but the actual growth depends also on the environment, which is expressed by rewriting the per capita growth equation (10.1) in terms of $E_j(t)$ and $C(t)$ as

$$r_j(t) = E_j(t)[\delta - C(t)] - \delta. \qquad (10.4)$$

Because $C(t)$ occurs in this equation multiplied by the environmental response, $E_j(t)$, the effect that competition of a given magnitude has on population growth depends on the environment. If the environment is not favorable for species j, that is, $E_j(t)$ is small, then the effect of $C(t)$ is small. Indeed, in the extreme case where $E_j(t) = 0$, $C(t)$ has no effect on $r_j(t)$, which then is fixed at the negative value $-\delta$. Similarly, if competition is strong, for example, near its maximum value $C(t) = \delta$, the effect of $E_j(t)$ on population growth is reduced. This effect, where an unfavorable value of one factor (either environment or competition) limits the impact of the other factor, especially the other factor's unfavorable impacts, may be thought of as buffered population growth. For this particular model, the buffering effect results in a minimum per

219

capita growth rate of $-\delta$ regardless of how unfavorable environmental and competitive conditions may be.

Various factors can modify the buffering effect embodied in equation (10.4). For example, it would be strengthened if the environmental response also modulated respiration. In the most extreme form, respiration would be proportional to $E_j(t)$, then δ would not be the sum of μ and γ, but simply μ and $r_i(t)$ would take the value $-\mu$ when $E_j(t) = 0$, although this no longer defines a minimum value for $r_j(t)$ under all conditions. In statistical terms, buffering, which is referred to also as *subadditivity* (Chesson 1994), comes from the interaction between environment and competition. The interaction determines how $r_j(t)$ changes as a result of joint changes in $E_j(t)$ and $C(t)$. The concept of covariance between environment and competition concerns how $E_j(t)$ and $C(t)$ do change jointly. These concepts are distinct, but the importance of each depends on the other. In particular, the product form of (10.4) that leads to the buffering effect also means that the average of $r_j(t)$ over time, $\bar{r}_j$, takes the following form:

$$\bar{r}_j = \bar{E}_j(\delta - \bar{C}) - \delta - \mathrm{Cov}(E_j, C) \qquad (10.5)$$

that is, the long-term growth of the population depends on covariance over time of the environmental response and competition (Chesson & Huntly 1997).

To see that species-specific responses to the environment, buffered population growth, and covariance between environment and competition jointly act to promote species coexistence in a variable environment, we first consider dynamics in a constant environment. In the absence of environmental variation, that is, if the $E_j(t)$ were constant over time, and the dynamics of resources allowed an equilibrium to occur, only one species would persist at that equilibrium and it would be the species with the largest value of E_j. That species would drive $C(t)$ to the value

$$C_j^* = \delta\left(1 - \frac{1}{E_j}\right) \qquad (10.6)$$

at which the r's of all other species are negative, meaning that these other species would go extinct. We can think of this result as a C^* rule for species limited in the same way by common limiting resources. It is a simple generalization of the R^* rule (Holt, Grover, and Tilman 1994; Tilman 1990a,b) for species limited by the same resource. Note that, in this case, the value of competition, $C(t)$, is a function of the environmental response of the dominant species.

When the $E_j(t)$ fluctuate over time, $C(t)$ must fluctuate over time too. In general, it should be expected that the fluctuations in $C(t)$ would be correlated with those in $E_j(t)$, at least if species j is abundant enough to have much competitive effect. For example, if species j is the only abundant species, and $E_j(t)$ varies only slowly over time, then $C(t)$ ought to track the value given by (10.6), that is, it would be an increasing function of $E_j(t)$, so that the covariance, $\text{Cov}(E_j, C)$, would be positive, and therefore would decrease the value of $\bar{r}_j$ in the long-term growth equation (10.5). As discussed in detail for general models (Chesson 1994, Chesson and Huntly 1997), species at high density tend to have positive values of $\text{Cov}(E_j, C)$, reducing their rates of increase over those predicted by the averages of E and C alone. Species at low density tend to have low values of the covariance if their environmental responses fluctuate asynchronously with those of their higher-density competitors. Thus, these low-density species have a growth rate advantage and tend to increase to higher densities. The presence of the covariance in formula (10.5) therefore tends to keep species in the system. For a single-species system, the magnitude of the covariance is approximately proportional to the variance, σ^2, of the natural log of E_j. This means that in a two-species system, a species at low density whose environmental response is uncorrelated with that of the other species would be advantaged relative to the other species by an amount proportional to σ^2. If it is negatively correlated with the other species, its advantage would be more; if positively correlated, its advantage would be less. The advantage that accrues to a species at low density tends to decrease with the number of species in the system, making coexistence more difficult and decreasing rates of recovery from low density (Chesson 1994).

In summary, plant species growing and interacting through resources according to the common ecosystem model are potentially able to coexist by the storage effect when they have different temporal patterns for their environmental responses. The three ingredients of the storage effect are automatically present in this model. Coexistence is not guaranteed, however, because like most mechanisms of coexistence, the storage effect can be overcome by large average fitness differences between species (Chesson 2000b). In particular cases, formulae are available to determine the fitness differences compatible with coexistence (Chesson 1994, 2000a). Particular examples of coexistence by the storage effect are given in particular versions of the common ecosystem model below.

Spatial and Spatio-Temporal Niches

Many different models over a long period of time have demonstrated species coexistence as a result of spatial and spatio-temporal niches (Levin 1974; Shmida and Ellner 1984; Chesson 1985; Comins and Noble 1985; Iwasa and Roughgarden 1986; Muko and Iwasa 2000). The formal development in terms of the storage effect is recent, but again reveals three key requirements: species-specific responses to the environment, covariance between environment and competition, and buffered population growth (Chesson 2000a). In the spatial domain, buffered population growth arises essentially automatically from the dispersion of populations in space, but the magnitude of the buffering is affected by the nature of dispersal and the nature of the environmental variation. In important cases, buffering in space may be shown to be equal to the maximum attainable buffering in time (Chesson 1984, 1985, 2000a). Thus, buffering in space may often exceed buffering in time, although not necessarily by very much.

Covariance between environment and competition in space arises in many circumstances, but is not automatic; for example, pure spatio-temporal variation in mortality rates during periods when species are dormant, and not actively drawing on resources, does not lead to covariance between environment and competition and does not promote coexistence (Chesson 1985). Pure spatial variation in such mortality rates, however, does lead to covariance between environment and competition under restricted dispersal scenarios, because then population densities increase at favorable locations, producing more competition (Chesson 2000a). Species-specific responses to spatially varying environments seem widespread in nature, which means there are many situations in which spatial niches of some form are a factor in coexistence. Of most importance for the question of ecosystem functioning is the way in which spatial variation may maintain a variety of species adapted to specific environmental conditions.

ECOSYSTEM FUNCTIONING

Although there are many results on species coexistence by means of environmental niches, as discussed above, information needed to understand ecosystem functioning is much more limited. We first consider simple situations that can be solved analytically for the cases of spatial and temporal niches, and then we move on to look at a more complicated model of a Mediterranean-type ecosystem by means of simulation.

Ecosystem Functioning with Spatial Niches

We suppose here that the environment varies in space, and that in equation (10.1) $E_j(t)$ takes the form $E_j(t) = f(x_j, E)$ where E is a quantity defining the physical environment for a given locality in space. The sole quantity distinguishing species j from other species is x_j, which defines the optimal physical environment for species j. The function f converts the distance α_j between x_j and E into $E_j(t)$, the growth activity of species j in an environment with the value E. We assume that f is nonincreasing and has the value 1 for a perfect match of species and environment, that is, $f(E, E) = 1$, which we may do without loss of generality.

Now consider a spatial network of a large number of local communities, each with a different value of E, and each large enough in area that interpatch dispersal has a negligible effect on local dynamics (except during initial colonization). For simplicity, we randomly assign a value of E to each local community in the network by drawing E's from a uniform probability density on an interval of length Δ_E. One may think of Δ_E as the regional range of environmental conditions. Similarly, we produce a regional flora of species available for colonizing the network by drawing values of x_j from a Poisson process on the E axis, with intensity λ. In other words, we assume that the x's "rain" down upon the E axis entirely at random, with an average of λ species per unit-length of E. Although these uniformity assumptions may seem restrictive, they can be greatly relaxed without altering the qualitative outcome.

The main quantities of interest are (1) the average number of species that will coexist in a regional network and (2) the average ecosystem function. Because $f(x_j, E)$ contains the only species-specific parameter in the model, the species from the regional pool with the highest value of $f(x_j, E)$ for a particular locality will competitively exclude all others from the pool in that locality. Because of the large number of local communities, every species in the regional flora with a niche position (value of x_j) contained in Δ_E must persist because it will dominate at least one local community. On average, there will be $\lambda\Delta_E$ such species. There may be up to two additional species that persist (the species with niche positions just before and just after the interval Δ_E), but the exact expression for the expected total number of species at equilibrium (D^*) is cumbersome and so we will use the close approximation $D^* \approx \lambda\Delta_E$.

Also, rather than calculate the network-average separately for each ecosystem function, we instead calculate only f^*, the expectation of $f(x_j, E)$ at equilibrium:

$$f^* = \int\limits_{\alpha=0}^{\infty} 2\lambda e^{-2\alpha\lambda} f(E + \alpha, E) \, d\alpha.$$

Average values of net primary production, carbon storage, N-mineralization, and evapotranspiration are all simple increasing functions of f^* under the particular circumstances assumed here. The expression for f^* relies on the fact that the probability that no species is present in the regional flora with a value of x_j within a distance α of any particular value of E is $e^{-2\alpha\lambda}$ (the zero term from a Poisson distribution).

Suppose $f(x, E)$ is given by the exponential function: $e^{-2|x - E|/\omega}$, where ω is the niche width (f is larger than $1/e$ over a range of environmental conditions of width ω). Then, using the fact that $\lambda = D^*/\Delta_E$, we find that f^* is simply $\theta/(1 + \theta)$, where $\theta = D^*\omega/\Delta_E$. Similarly, if $f(x, E)$ is approximately a step function, nearly equal to 1 over the interval $x - \omega/2 \leq E \leq x + \omega/2$, but highest at $x = E$, and equal to zero outside this interval, then approximately: $f^* = 1 - e^{-\theta}$.

These two functions for f^* are almost identical. Each shows that ecosystem function increases with diversity, initially with slope ω/Δ_E, and asymptotes at $f^* = 1$ for $D^* \gg \Delta_E/\omega$. The critical point is that the diversity necessary to maintain ecosystem function is set by each species' capacity to buffer environmental variation (ω) and by the magnitude (Δ_E) of environmental variation present. The necessary level of diversity becomes arbitrarily large as we expand the range of environmental conditions present in the region.

Ecosystem Functioning with Temporal Niches: Lottery Models

Suppose that the habitat is divided into cells, and that every occupied cell contains the same total biomass: B_{max}. We assume discrete time and proceed from one year to the next in three steps. First, random mortality occurs, killing the biomass in each cell with probability μ. Second, analogous to the assignment of environments in the spatial model above, we choose an environmental condition E for the year from a uniform distribution on the interval Δ_E. This is done independently each year. Third, we choose the second function f from the spatial examples above for which $f(x_j, E) > 0$ only over a finite interval ($x_j - \omega/2 \leq E \leq x_j + \omega/2$), and with its maximum at $E = x_j$. We say that a species-j's niche *contains* a value of E if $f(x_j, E)$ is nonzero for that value. Species are assigned environmental niches exactly identically to the spatial problems as a Poisson process on the E axis. Only species whose environmental niches contain the value of E for that year reproduce and no reproduction occurs if the environmental condition is outside every species' niche. If reproduction occurs, new recruits fill all empty space and the species identities of the new recruits are determined by lottery competition (Chesson and Warner 1981). The lottery may range from completely biased in favor of the species with the largest value of $f(x_j,E)$, so that its offspring capture all empty sites, to unbiased, so that reproducing species capture sites in proportion to their abundances.

The long-term mean and variance of total community biomass produced by this stochastic process is derived in the appendix as

$$\text{mean} = \frac{1-q}{q\mu} B_{\max} \ln\left(1 + \frac{q\mu}{1-q}\right)$$

$$\text{variance} = \frac{1-q}{q} \frac{B^2_{\max}}{1-(1-\mu)^2} \ln\left(\frac{1-q(1-\mu)^2}{1-q}\right) - (\text{mean})^2$$

where q is the fraction of the interval Δ_E not included within the niche of any of the coexisting species. Because a sufficient condition for the indefinite persistence of a species is that it is the only species that reproduces in some years (that is, a portion of the species' niche is contained within Δ_E but does not overlap the niche of any other species), it is straightforward to show that $q = e^{-\omega\lambda}$.

Although we do not have a general formula for the number of species that coexist, we can produce some results for the bracketing extremes of a completely biased lottery (the species with the largest nonzero f captures all vacant sites in that year) and an unbiased lottery. The purely biased lottery is exactly like the spatial model in that all species with values of x_j contained by Δ_E will coexist, plus up to two others with niche positions just outside Δ_E. We thus can use the close approximation $D^* = \lambda\Delta_E$, for the average number of coexisting species. In the case in which q is small, so that terms of order q^2 or higher are negligible, we find that the mean total biomass reduces to $1 - e^{-\theta}$, where $\theta = D^*\omega/\Delta_E$. This is exactly the same expression obtained for the analogous spatial model. Also, the coefficient of variation squared of total biomass (variance/mean^2) is simply $e^{-\theta}$, showing that the temporal stability of ecosystem functioning also increases with diversity. This effect of diversity on stability is a direct consequence of partitioning of temporal heterogeneity, which is also responsible for the effect of diversity on average functioning.

Finally, in the case of a purely unbiased lottery, we can produce bounds for the amount of diversity that will be maintained in the limit of a large pool of potential species (the limit of large λ). These bounds set the level of diversity necessary to maximize both functioning (the amount of total biomass) and the stability of functioning. A necessary condition for indefinite persistence under our unbiased lottery is that each coexisting species is the only species that reproduces in some fraction of years. We thus obtain bounds on the number of coexisting species as the minimum and maximum number of line segments of length ω that can be placed along the E axis such that at least one portion of each segment both overlaps Δ_E and does not overlap any other line segment:

$$\frac{\Delta_E}{\omega} \leq D^* < 2\,\frac{\Delta_E}{\omega} + 2.$$

As in all of the other cases of coexistence caused by spatial or temporal heterogeneity, the diversity necessary to maximize functioning increases without bound as the capacity of each species to buffer environmental change decreases and as the amount of environmental heterogeneity increases.

Ecosystem Functioning with Temporal Niches: A Mediterranean Ecosystem

To examine ecosystem functioning in a more detailed, though less general, way than has been possible with the analytical approaches above, we use simulations of equations (10.1) and (10.2) for a roughly Mediterranean-type climate (Cody and Mooney 1978; Hooper and Vitousek 1997). Thus, rainfall has a substantial peak in midwinter (figure 10.1). Rather than an extreme Mediterranean pattern with very little rainfall at other times of the year, rainfall consists of stochastic variation about a sinusoidal pattern. In contrast to the analytical approach taken above, and in contrast to most studies of the storage effect, the environmental re-

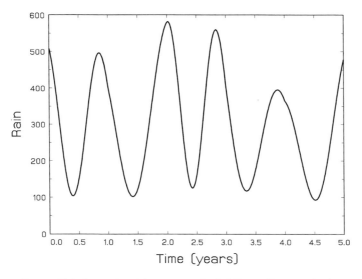

FIGURE 10.1. A sample of the pattern of rainfall over five years in the simulations for the base parameter set. Integer values of time coincide with mean peak rainfall; mean minimum rainfall is midway between integer values. Units are mm/year. Table 10.1 explains the generation of this pattern.

sponses of the different species have been chosen to be deterministic—for example, cued to photoperiod (figure 10.2). Thus, the temporal niches that these species exhibit are seasonal niches. The environmental responses thus peak at different times for different species.

The amplitudes of the environmental responses vary from species to species depending on the time of year at which the environmental response peaks (figure 10.2), which is intended to mimic the effect of temperature on plant growth with warmer conditions promoting stronger growth for given levels of resource availability. Such amplitude differences also provide a trade-off, permitting a species that grows at a disadvantageous time with respect to rainfall to compete successfully with other species. These deterministic environmental responses are chosen to emphasize that the

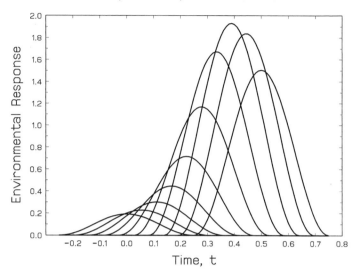

FIGURE 10.2. Environmental responses, $E_j(t)$, of the 10 species in the global species pool. Each response is a complete cycle of a sine wave over a six-month period, and is zero outside this period. The change in peak height of $E_j(t)$ with the location of the peak (phase) defines a trade-off promoting coexistence of these species. Time t is time in fractions of a year, and 0 defines the mean time of peak rainfall for the unperturbed system.

storage effect works both with deterministic as well as stochastic environmental responses. They provide results that complement those on stochastic environmental responses in the analytical section and in simulations published elsewhere (Lehman and Tilman 2000).

Though the niches of the species are deterministic, their growth is far from deterministic as they are affected greatly by stochastic fluctuations in resources. The presence of different species coexisting because they have different temporal niches might be expected to lead to less temporal variation in resource levels, and, indeed, lower average resource levels due to fluctuations in resource supply. Higher average primary productivity might also be expected. To examine these possibilities, rainfall in this system varied stochastically in several ways from year to year: total amount of rain, magnitude of seasonal fluctuation, and the phasing of the winter peak in rainfall. As rainfall affects nutrient recycling, variation in rainfall should also cause fluctuations in nitrogen availability, which might also be dampened in the presence of species coexisting with temporal niches. The nature of these temporal niches, including the fraction of respira-

Table 10.1. Parameters of the Simulation Model

Parameter Description	Parameter Symbol	Value
Reference value of r	a	1077
Respiration loss	γ	0.1
Tissue death	μ	0.1
Light half saturation	K_L	50
Nitrogen half saturation	K_N	0.005
Water half saturation	K_W	200
Water use efficiency	U	0.001
Litter fraction to fast carbon pool	f	0.5
Water loss from root zone	$s + \epsilon$	0.55
Nitrogen leaching	ζ	0.2667
Nitrogen deposition	D	0.004
Effect of water availability on decomposition	ϕ	0.00002
Fast C pool decomposition	λ_f	1
Nitrogen-to-carbon ratio of slow carbon pool	ν_S	0.00444
Nitrogen-to-carbon ratio of fast carbon pool	ν_f	0.06666

Notes: The rainfall pattern was generated by the formula

$$R_{t+u} = a_{t+u} + b_{t+u}\{1 + \cos[2\pi(u - \theta_{t+u})]\},$$

where R_{t+u} is the rain intensity at time u ($0 < u < 1$), in year t, in units of mm per annum; for integer t, a_t and b_t are lognormal random variables with means 100 (150 for higher rain) and 250, respectively, with standard deviations of their natural logs equal to 0.25; θ_t is normal with mean zero (0.1 for later rain) and standard deviation 0.2; and $a_{t+u} = (1 - u)a_t + ua_{t+1}$, with corresponding definitions for b_{t+u} and θ_{t+u}.

tion and tissue death that varies in proportion with growth activity, is likely to have important effects on such outcomes. Table 10.1 gives the parameters of this system.

For these simulations, 10 species were selected that are capable of long-term coexistence in the system, as determined by their persistence together over 2000 years. These species are defined by their environmental responses as depicted in figure 10.2. Simulations were then done by selecting local species pools varying in size from one to nine species from this 10-species global pool. For local pools of sizes one and nine there are 10 possible selections, each of which was chosen. For other local pool sizes, 20 random selections from the global pool were used. Local species pools were treated as initial local communities, with species given equal biomasses summing to the equilibrium biomass for the system applicable in the absence of any temporal environmental variation. Data for the first 1000 years were discarded to remove the effect of the arbitrariness of the initial conditions. Competitive exclusions occurred during this initial period and data were collected for the subsequent 100 years on the community surviving from the local species pool.

Three different environmental conditions were considered. A baseline condition was 350 mm of rain per year with mean phase of zero, corresponding to the phase of the environmental response of the earliest species from the species pool. The two perturbations to this condition, which were treated separately, were (a) 400 mm per year, and (b) a shift in phase to 0.1, that is, a delay in rainfall of one-tenth of a year. As the species are distinguished by the phases and amplitudes of their environmental responses, rather than by their responses to the amount of rainfall, these perturbations correspond to (a) an environmental change orthogonal to the niche axis, and (b) a change aligned with the niche axis. Simulations were done using the Simgauss module of the Gauss Mathematical and Statistical System (Aptech Systems, Inc. Maple Valley, Washington, USA).

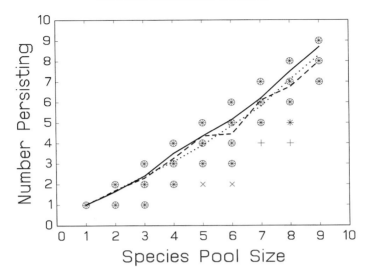

FIGURE 10.3. Persisting species for each local pool size. O: unperturbed; +: higher rain; ×: later rain. Means are given by solid, dotted, and dashed lines, respectively.

SIMULATION RESULTS: PERSISTENCE AND COEXISTENCE

The species always persisted in a long-term, stable fashion in all single-species runs. With more than one species, competitive exclusion sometimes occurred, but the average number of coexisting species was in general only about one less than the pool size (figure 10.3). For the purposes of determining the number of coexisting species, a species was considered persistent if its average biomass in the last 100 years of the simulation was at least 1% of the total biomass in the system. In general, species regarded as coexisting by this definition persisted with stable fluctuations, as depicted in figure 10.4.

MEAN ECOSYSTEM FUNCTIONING

Mean ecosystem functioning increased with the size of the local species pool, up to an asymptote achieved by approximately six species, depending just a little on the partic-

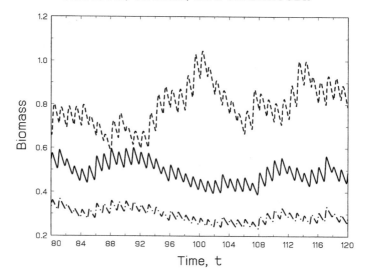

FIGURE 10.4. Sample of a simulation run giving biomasses of a three-species community. The bottom, middle, and top curves are the first, fifth, and tenth species, respectively, of the global species pool labeled in order of phase of their environmental responses (see figure 10.2).

ular ecological process chosen (figure 10.5): There is a clear monotonic increase in maximum ecosystem functioning at each pool size from one to three selected species in every case, after which these maxima are approximately constant. The two perturbations had dramatically different effects. Increase in the amount of rainfall substantially increased ecosystem functioning to a similar degree for every species pool size. A delay in rainfall also uniformly increased mean ecosystem functioning, but to a much lesser extent than the increase in rainfall. If these data are plotted against the number of coexisting species, rather than against the species pool, the picture changes a little: most features of these graphs are shifted toward the origin by one species, as one might expect from the fact that roughly one species went extinct on average (figure 10.3).

234

Mean Ecosystem Functioning

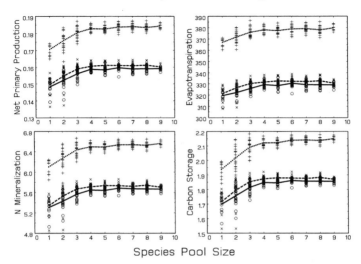

Species Pool Size

FIGURE 10.5. Means of ecosystem services for communities arising from randomly chosen local species pools of different sizes. Each point is the mean of five replicates for a particular local species pool and environmental regime. ○: unperturbed; +: higher rain; ×: later rain. Means are given by solid, dotted, and dashed lines, respectively. Units are $kg/m^2/year$ (net primary production), $mm/year$ (evapotranspiration), $g/m^2/year$ (N-mineralization), and $kg/m^2/year$ (carbon storage).

VARIANCE OF ECOSYSTEM FUNCTIONING

Variation over time was measured continuously, including seasonal as well as year-to-year components of variation. So that the mean did not get confounded with variation, relative variation in all quantities was considered. For a positive quantity, the standard deviation of the natural log of the quantity is a good measure of relative variation, and is the preferred measure here because its standard error can be calculated very easily over replicate runs. Net primary productivity, although having a positive average in persistent vegetation, has periods when it is negative, precluding the use of the standard deviation of the natural log. The coeffi-

Variability of Ecosystem Functioning

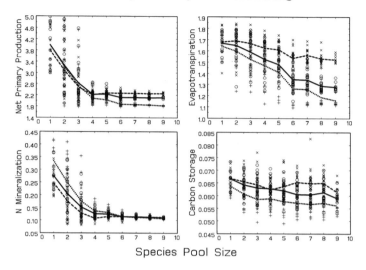

Species Pool Size

FIGURE 10.6. Variability of ecosystem services defined as coefficient of variation over time for net primary production, and standard deviation of the log over time for the other ecosystem services, which always take positive values. Symbols are as given in figure 10.5.

cient of variation, which has similar properties to the standard deviation of the natural log, was used instead. When both measures were available, they were found to give nearly identical results. These measures are equivalent to the reciprocal of the stability measure of Lehman and Tilman (2000).

Variability of ecosystem functioning declined with local pool size for all ecosystem functions, generally achieving an approximate minimum value with a pool size of six species (figure 10.6). However, for evapotranspiration, variability continues to decrease for local pool sizes above six. Indeed, at least for 400 mm of rain, it appears to decline over the entire range, 1–9, of local species pools. In these simulations, water availability was set up to be the limiting factor, and rain was a variable input. Having more species potentially gives more even use of water through the year, reducing fluctuations in soil moisture.

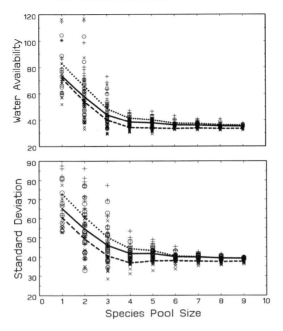

FIGURE 10.7. Soil water content (mm) and temporal standard deviation of water content for the simulations. Symbols are as given in figure 10.5.

The perturbations had similar effects on variability for most ecosystem functions, with later rain generally giving the most variability and more rain, the least variability. A notable exception is nitrogen mineralization where this pattern is reversed, presumably explained by higher soil moisture, and higher soil moisture variability at higher rain (figure 10.7) coupled with the strongly nonlinear relationship between soil moisture and nitrogen mineralization assumed in the common ecosystem model.

DISCUSSION

The general prediction that ecosystem functioning, especially primary production, should increase with the number of species was borne out. At the simplest level, this result can be understood as the filling of niche space. With niches defined temporally, this could mean simply having species present, actively photosynthesizing and consuming resources for the maximum amount of time. In this way, the species present are complementary to one another, which leads to the phenomenon of overyielding: the combined output of the community is higher than that of the most productive species when present alone (Hector 1998; Hector et al. 1999; Tilman 1999b). This result is illustrated by our simulations, which show that the best monocultures are inferior to some mixed communities (figure 10.5).

Two alternatives to niche complementarity have been discussed in the literature: positive species interactions (Hector et al. 1999) and the sampling effect (Huston 1997; Tilman, Lehman, and Thomson 1997; Loreau 1998b; Tilman 1999b; Hector, Chapter 4 herein). Positive species interactions require one or more species to directly benefit other species, for example, by facilitating nitrogen fixation or increasing humidity, situations not accommodated in our models but that may well be important in nature (Hector et al. 1999). Under the sampling hypothesis, the more species there are, the better the performance of the best species is likely to be simply because the maximum of a sample tends to increase with the sample size (Tilman 1999b). Under the sampling hypothesis, the best species dominates and is responsible for the high productivity. The sampling and complementary niche hypotheses appear to be distinguished by distinct predictions. With sampling, the best multispecies community cannot be better than the best single-species community. The best multispecies community can only equal the best single-species community by being dominated by the best species for that locality. In contrast, the complementary niche hypothesis predicts that the best multispecies system will have higher productivity than the best single-species system.

Although sampling and complementary niches do appear distinct in homogeneous environments, in heterogeneous environments that distinction is lost (Crawley et al. 1999).

For example, consider the simple model of spatial niches discussed above. In each patch, sampling prevails: the best species to arrive there dominates. However, if several patches with different environments are considered as a unit, they will likely have different species dominating them and the unit will have several species coexisting. The over-yielding phenomenon is predicted because no one of them is superior over all environments; the productivity of the several-patch system is higher than it would be under its best monoculture; and individual species have higher productivity per unit biomass because they are concentrated in localities where their performance is better.

The corresponding temporal phenomenon is illustrated above in the lottery form of community dynamics in a temporally variable environment. In the "completely biased" version, there is a single winner in competition for any particular year. That species gains all newly available sites. However, due to the simplified assumptions of that model, there are only two possibilities for productivity: production occurs by capture of newly available sites and either all of these are taken or none of them is taken. For a given year, the model satisfies a particular version of the sampling hypothesis: either a species present is adapted to the environment of the year, and all new sites are captured by new recruits, or no species is adapted to the year, and no new sites are captured. More species increase the probability that all sites are captured, but the productivity of the year is no better than a monoculture of the best species for the environmental conditions in that year. When productivity is combined over several years with different environmental conditions, however, the best monoculture is not as good as the best mixture, because the best monoculture will likely miss capturing space in some years. The more species there are, the more years have some adapted species, and the higher the productivity is over those years. Thus, overyielding is seen for the community as a whole for productivity measured over a sequence of years.

The common ecosystem model in the form given by equations (10.1) and (10.2) also contains elements of the sampling mechanism, which are most apparent when the environment changes slowly relative to population growth rates. The most productive species for a particular environmental state would tend to be dominant, driving other species to low density. Thus, the system would be dominated most of the time by the most productive species for the environment of the time, assuming of course that the conditions for species coexistence are satisfied and all species do remain in the system. In the short-term, the predictions of the sampling hypothesis prevail, but again, when averaged over time, the predictions of complementary niches prevail.

When the rate of population growth is more comparable to the rate of environmental change, a mixture of species will be present at any one time, which will most likely mean that productivity at any given time is lower than in cases where the best species for that time comes to dominate. For example, in the simulated Mediterranean system, turnover of biomass was low because tissue death was only 0.1 per unit biomass per year, while species' activity functions go through their full cycle within a year. Thus, species do not greatly change their biomass within a year, even though their productivity undergoes dramatic changes. One might ask how close to the full effect of sampling this could be. Although we cannot fully answer this question, it is clear that for temporal niches to achieve the same effect as sampling at any given time, species would have to be completely inactive, and hence not drawing on water and nitrogen when they are not the best. During this time also, they should not suffer respiratory losses or tissue death and should not block the light to other species (light limitation is minimal or species are deciduous). Deviations from any of these conditions will reduce productivity. Thus, the strongest performance of ecosystems where species coexist by temporal niches is predicted to occur when species go dormant at times when they are not dominant, and are able to build up quickly in size and number when they are dominant. Annual species and perennials that die back or are deciduous at their disfavored times add some of these elements to a community, but a system consisting of species that are dormant whenever not dominant exists only as theoretical fantasy.

Spatial and spatio-temporal niches share the property with temporal niches that mean ecosystem functioning is lower than would obtain if for any place and time, the best species for that time and place were present alone at its equilibrium density. In general, a given time and place will have species in addition to the best species for the environmental conditions there. However, a close match to the maximum set by the sampling process is possible with pure spatial variation as the best species at a site will tend to dominate there if dispersal is low (Chesson 2000a). Indeed, one might think of environmental niches as an elaboration of the sampling hypothesis, adding to it the mechanism of maintenance of regional diversity on which it depends, and which is omitted from general descriptions of the hypothesis. Where the sampling hypothesis has been examined empirically (Hector et al. 1999), there is no expectation that the best species on a plot would fully displace other species, and so in practice productivity would never equal that of the best monoculture. Thus, in practice, the sampling hypothesis does not achieve the theoretical sampling maximum. The truth is that the sampling and niche complementarity hypotheses are not distinct. If it is accepted that differential performance in different environments defines niche differences, as in our definition of environmental niches, niche complementarity and sampling in a variable environment simply cannot be thought of as different phenomena. Indeed, complementary environmental niches allow the sampling mechanism to occur locally in time and space, but its original predictions will not be borne out when the sampling unit includes ecologically distinct environments, either spatially or temporally. Instead, the predictions of niche complementarity will prevail.

The purported properties of the sampling process also depend on the postulate that the best competitor is the most productive (Loreau 1998b), an assumption that is sometimes violated in nature (Hooper and Vitousek 1997). This requirement follows from the assumptions made here that species differ only in their environmental responses as defined in equation (10.1). If species differ in water use efficiency or in the ratio of respiration to tissue death, this assumption can be violated. In principle, it would be possible for net primary productivity to decline with the size of the species pool if water use efficiency declined sharply with competitive ability or respiratory losses increased.

Reduction in variance with number of species has also been found in several different theoretical studies (Doak et al. 1998; Tilman, Lehman, and Thomson 1998; Ives, Gross, and Klug 1999; Tilman 1999b; Yachi and Loreau 1999) and in empirical studies of experimentally constructed communities (Tilman 1999b). Ives, Gross, and Klug (1999) make the point that the species in a community must be diversified in their responses to environmental conditions for such results to occur, but they provide no mechanism leading to this outcome. Temporal niches provide this mechanism. The requirement of species-specific responses to the environment means that coexisting species are necessarily diversified in their environmental responses. The species coexist because of their complementary use of resources over time. Species can genuinely be seen as taking advantage of temporally arising opportunities. A diverse species pool improves the probability of a good match between species and the available environmental conditions, and leads to reduction in variance with diversity over the full range of ecosystem functions.

A system diversified with respect to responses to environmental conditions would also be expected to show maintenance of ecosystem functioning in the face of environmental change. The simulations here considered two different sorts of environmental change of comparable magnitude, but of very different nature. Delaying the peak rainfall by a tenth of a year increased mean ecosystem functioning, but by less than 5% in general. As the system is diversified along the seasonal axis, the smallness of this effect is not surprising. For a given set of coexisting species, relative abundances shift, reflecting the change in conditions, emphasizing species more suited to the new conditions (figure 10.8), and therefore maintaining ecosystem functioning. More surprisingly, this delay in rainfall was accompanied by higher variability in ecosystem functioning, perhaps reflecting the greater probability that rainfall would occur when species were not so available to make use of it.

Increasing rainfall by 50 mm per year had a much greater effect than a change in timing. As water was the limiting resource, this is hardly surprising. Being orthogonal to the niche axis, it is one that would be expected to affect all species equally. Consistent with this prediction is the absence of appreciable change in the mean phase of the environmental response for communities experiencing this perturbation (figure 10.8). The decreased temporal variation with increased rainfall is to be expected as the ratio of winter to summer rain decreased with the increase in annual rainfall, likely decreasing seasonal differences in productivity.

With changes along an axis of differentiation, such as timing of rainfall, presence in the species pool of species adapted to the new conditions is critical. Such species may exist in low abundance, maintained perhaps by the tails of the distribution of environmental events for the unperturbed situation. Those rare species may be the mainstay of a system after perturbation. The same may be true for space. Parts of the habitat that are seen as extreme may become the norm—and species dominating there, although rare because their habitat is rare, may become the mainstay following an environmental change. Indeed, such rare species may provide much of the resistance to perturbation for a

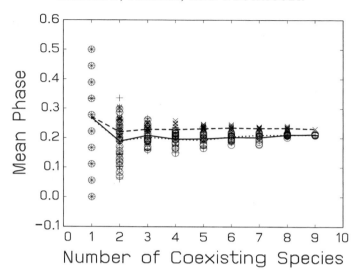

FIGURE 10.8. Mean phase of the environmental response in each simulated community for each environmental regime. (Mean phase is calculated as the average of the phases of the constituent species weighted by their mean biomasses.)

system's ecosystem functions. These considerations lead to the prediction that the maintenance of ecosystem functioning in the presence of long-term change should decline with declining diversity. There is little evidence of this phenomenon in the results of the simulation studies, perhaps because of the modest nature of the perturbation of the environment compared with the niche breadths of the species. This result is to be contrasted with the major effect of diversity on variance in ecosystem functioning in the presence of seasonal and year-to-year environmental variation, and the important effects of diversity on mean ecosystem functioning found here in association with environmental niches.

ACKNOWLEDGMENTS

We are especially grateful to Ann Kinzig for guidance in parameterizing the simulation model, and to Clarence Lehman for helpful discussion of the topic. P.C. was supported by NSF grant 9981926.

APPENDIX

To obtain the mean and variance expressions for the lottery model, we begin by setting total biomass density $(B(t))$ at B_{max}. If T is the first time the environmental condition (E) hits a value contained by a species' niche, then $B(T) = B_{max}$ and $B(k) = B_{max}(1 - \mu)^{k-1}$, for $k = 1, 2, \ldots,$ $T - 1$. At time T, we reset time to zero and repeat the process. The distribution of T is geometric with $\text{Prob}(T = k) = q^{k-1}(1 - q)$, where q is the fraction of Δ_E not contained by any species' niche. We define $B([0, m - 1])$ as

$$B([0, m - 1]) = \frac{1}{m} \sum_{i=0}^{m} B(i).$$

Then, the long-term average biomass is the expectation of $B([0, T - 1])$, which is calculated as the sum over all possible values of k of $\text{Prob}(T = k)$ times $B([0, k - 1])$:

$$\sum_{k=1}^{\infty} q^{k-1} (1 - q) \frac{1}{k} \sum_{i=0}^{k-1} (1 - \mu)^i B_{max}$$

which reduces after some algebra to the mean biomass given in the text. Similarly, the expectation of $B([0, T - 1])^2$ is given by

$$\sum_{k=1}^{\infty} q^{k-1} (1 - q) \frac{1}{k} \sum_{i=0}^{k-1} [(1 - \mu)^i B_{max}]^2.$$

After subtracting the square of the mean biomass, this reduces to the expression for the variance of biomass that is given in the text.

245

Biodiversity and Ecosystem Functioning: The Role of Trophic Interactions and the Importance of System Openness

Robert D. Holt and Michel Loreau

INTRODUCTION

A central theme in ecology is that population dynamics, species coexistence, and, ultimately, the entire organization of communities are all profoundly influenced by the complex web of trophic interactions that binds the lives of species together (Pimm 1982; Polis and Winemiller 1996). (The term "trophic interaction" here denotes feeding relationships between species, usually implying transfers of energy and nutrients.) A priori, given the growing evidence for trophic cascades (Pace et al. 1999) and other system-wide manifestations of trophic interactions (e.g., Elliott et al. 1983, Hairston and Hairston 1993; DeRuiter, Neutel, and Moore 1995; Grover and Loreau 1996), it would be shocking if trophic interactions did not significantly influence the impact of biodiversity on ecosystem processes. Indeed, there is suggestive evidence in the literature that the detailed character of trophic interactions has strong ecosystem impacts (Pastor and Cohen 1997; Naeem and Li 1998). The presence or absence of single consumer species or trophic guilds can have large ecosystem effects (Huntly 1991; Jones and Lawton 1995; see, e.g., McNaughton, Banyikwa, and McNaughton 1997; Mulder et al. 1999). For instance, at Cedar Creek, deer exclosures led to substantial increases in both total plant biomass and soil nitrogen, concordant with changes in species composition (Ritchie, Tilman, and Krops 1998). Our purpose in this chapter is not to review such examples (which we find compelling) but to address conceptual issues on the interface of food web ecology and ecosystem ecology.

To extend the theory of plant diversity and ecosystem functioning (see chapters 2, 9, and 10) so as to encompass food web effects, we consider a two-trophic-level analogue of the basic one-level ecosystem model (see chapter 8), including explicit herbivore dynamics. Our basic approach will be to develop analogues of the "sampling effect" model (Tilman 1999b), where the sampling could occur either among producers facing herbivory, or instead among the herbivores themselves. Incorporating realistic food web dynamics greatly inflates the parameter set and analytical difficulty of the basic ecosystem model, and so as a "first pass" through the problem we sketch trends apparent from analyses of limiting cases of the basic model with herbivory. Because many consumers are highly mobile, relative to producers, it is natural to consider the consequences of different patterns of system "openness" when addressing the food web dimension of the diversity–ecosystem interface. After some general remarks on the relationship between the sampling model and community assembly, and the issues of trophic complexity and system openness, we describe how we incorporated herbivory into the basic model and present our findings. We then discuss important avenues for future work, and conclude by outlining a set of qualitative messages about the relationship of diversity and ecosystem functioning.

THE SAMPLING EFFECT MODEL
AND COMMUNITY ASSEMBLY

The "sampling effect" model relates diversity to ecosystem functioning by fusing two insights (Aarssen 1997; Tilman et al. 1997; Tilman 1999b; see also Huston 1997). The first is that the amount of interspecific phenotypic variation available that can impact ecosystem attributes should scale with the richness of the species pool from which local communities are assembled. The second is that local communities are restricted subsets of regional species pools (Roughgarden and Diamond 1986; Weiher and Keddy 1995), and that interspecific interactions restrict community membership. In some circumstances, simple rules determine which species drawn from the regional pool will dominate. For instance, the "R^*-rule" states that if a guild of competitors contends for a single resource in a stable environment, the species that persists at the lowest resource level displaces all others (Tilman 1982, 1990b).

We call such rules of local dominance and community membership "sorting rules." Tilman et al. (1997) use a sorting rule to show that total community biomass should increase monotonically with diversity, and that the pool of unconsumed resource should decline with diversity. Even complex systems at times match simple sorting rules (e.g., Tilman 1990b; Grover and Holt 1998). In this chapter we assume that species vary in such a way that a single species dominates in each trophic level. The basic question then is how rules of dominance (defining the outcome of species sorting at each level) map onto shifts in ecosystem processes, as subordinate species are replaced by dominant species during community assembly.

IMPORTANCE OF TROPHIC COMPLEXITY
AND SYSTEM OPENNESS

The phenotypic diversity of primary producers pales when compared with the diversity of heterotrophic consumers (see Naeem, chapter 5 herein). A substantial fraction of the diversity of life revolves around the diversity of ways heterotrophic consumers find, capture, and utilize their resources. This diversity of trophic interactions is expressed in the reticulate patterns of food webs and in the complexity of routes for nutrient recycling. Given the great variety of food web structures that exist among communities, in the search for general principles relating trophic interactions to ecosystem functioning we expect a priori to find numerous exceptions to any suggested generalization, depending on the detailed interaction structure of multitrophic communities. We suspect the great diversity of food web patterns may provide one class of explanations for variability among systems in ecosystem functioning.

For instance, consumers can profoundly change rules of coexistence and exclusion for producers, as well as alter the relative abundances of those producers that do coexist (Holt, Grover, and Tilman 1994; Grover and Holt 1998; Leibold 1996; Olff and Ritchie 1998). Take almost any model of competitive interactions among producers and add herbivory. If herbivores are tightly specialized to particular producer species, this facilitates coexistence. Generalist herbivores by contrast can either facilitate coexistence or make it more difficult (Grover and Holt 1998; Olff and Ritchie 1998). Moreover, herbivore effects on producer coexistence can be modulated by the action of higher trophic levels. Classic food chain models (Oksanen et al. 1981; Oksanen and Oksanen 2000) predict that the response of plant biomass to nutrient enrichment is profoundly impacted by interactions among higher trophic levels. But the precise relationship depends strongly on whether or not multiple species coexist in each trophic level, and the reticulate pattern of feeding links in the food web (Abrams 1993).

At times, single consumer species ("keystone" species) have disproportionately large effects on community structure. Theoretical and empirical studies of multispecies assemblages with strong interactions suggest that the impact of keystone species on ecosystem functioning depends on numerous details of both the biology of the keystone species and the interaction structure of the entire system. Counterintuitive effects frequently emerge because of the simultaneous interaction of multiple feedback loops and frequency-dependence in trophic interactions (Roughgarden 1976; Holt 1997).

Food web dynamics can also alter the diversity–ecosystem relationship by changing patterns of spatial coupling. Experimental field studies are conducted at small spatial scales where spatial flows can have a significant impact on ecosystem dynamics. There is a growing recognition of the importance of spatial flows among habitats as modulators of local food web and ecosystem dynamics (DeAngelis 1992; Holt 1996, 1997; Nisbet et al. 1997; Polis, Anderson, and Holt 1997; Huxel and McCann 1998; Loreau 1998a). At the basal level of abiotic resources in the basic ecosystem model (Pacala and Kinzig, chapter 8), rainfall and deposition provide inputs of abiotic resources in terrestrial habitats, and drainage and leaching likewise describe outputs. Consumers are often highly mobile and readily couple the dynamics of spatially distant communities (Holt 1996). Heterotrophic consumers thus provide multiple conduits for spatial exchange among ecosystems. System openness in general clouds rules of community structure derived from local exclusion (Loreau and Mouquet 1999; Holt and Gonzalez, ms.). Recurrent immigration can permit species to persist as community members, even though they tend toward local exclusion.

Ecosystems are likely to differ greatly in the way they are coupled to the external world via spatial flows at different trophic levels. We show below that the pattern of openness strongly influences how community rules of organization map onto ecosystem functioning (see also Holt, in press, and Loreau and Holt, in prep.).

TOWARD AN ECOSYSTEM MODEL
WITH TROPHIC INTERACTIONS

Modifying the basic ecosystem model of this book so as to capture the main chains of trophic interactions known to be significant in ecosystems (see Naeem, chapter 5) leads to models of daunting dimensionality and complexity. It is useful to consider first simpler systems. In the next few paragraphs, we describe how we splice herbivory into the basic ecosytem model (Pacala and Kinzig, chapter 8). The expanded model has two new compartments, describing herbivore population dynamics, and a detrital pool generated by the herbivore. We assume that herbivory is "laissez-faire" (Caughley and Lawton 1981); that the rate of per capita herbivory is given by a type II functional response, parameterized by an attack rate a', a saturation constant h, and an assimilation efficiency of c; and that herbivore deaths are constant (at per capita rate μ'). In a spatially open system with a single herbivore species, we assume a constant rate of immigration, I, and a constant per capita rate of emigration, e; in a spatially closed system, these parameters are set equal to zero. With these assumptions, the herbivore population has the following dynamics:

$$dH/dt = H[c\,a'B/\,(1\,+\,a'hB)\,-\mu'] + I - eH. \quad (11.1)$$

The herbivore generates a detrital pool of abundance C_H, which has the following dynamics:

$$dC_H/dt = H\mu' + H(1 - c)\,(a'B/(1 + a'hB)) - \lambda_f C_H. \quad (11.2)$$

We assume that herbivores potentially contribute to detrital pools via direct deaths, or unassimilated consumption, and do not discriminate between these two detrital sources. We also assume that decomposition is fast, at the same rate as the fast pool of producer-generated detritus. To complete the model, we must add a second mortality term in the producer equation of the basic model, describing mortality due to consumption by herbivores as follows: $-HBa'/(1 + a'hB)$. To simplify the algebra, we assume below that nutrient and water uptake are described by Michaelis-Menten kinetics, with $\theta = 1$. There is also an added term in the nitrogen equation.

Now imagine that a species pool exists for both herbivores and producers. As in simple one-trophic-level models of the sampling mechanism, we assume that the phenotypic variation in this pool is along a single axis, expressed as variation in a single rate parameter at a time. This constrained variation precludes local coexistence. However, as in the "sampling effect" model, regional species diversity could still influence local processes by increasing the range of variation in key system parameters available via colonization. We will allow colonization in just one trophic level at a time.

Case I: Ecosystem Closed at Top, Open at Bottom

Assume that a single producer species is present, and consider first species sorting of herbivores. With a single herbivore species, equilibrial plant biomass is $B^* = \mu'/a'$ $(c - \mu'h)$. When two herbivores are together in the same community, they will compete for resources (plant biomass). The winning species is the one that can persist at the lower B^*. The winning herbivore species will be the species with higher attack rate, a', or assimilation efficiency, c, or lower handling time, h, or direct density-independent mortality, μ'. Each parameter defines a different "sorting rule," or mechanistic reason for local competitive dominance by herbivores.

Now assume that a single herbivore species is present, and consider species sorting among plants, which vary in just one parameter of the model. A variant of the argument in Holt, Grover, and Tilman (1994) reveals that producer coexistence is impossible; the winning producer species is the one that both reduces resources to the lowest level, and supports the greatest herbivore numbers. In the general ecosystem model augmented with herbivory, this leads to seven "sorting rules": the dominant producer species will be the one with lower K_L, K_N, or K_W (greater resource uptake rates); or the one with higher r (greater maximal production); or the one with lower a' (lower vulnerability to herbivory); or with lower μ (losses other than to herbivory) or γ (respiratory costs).

Rather than lay out all the tedious algebraic details, table 11.1 describes the general direction of change in each of the ecosystem properties considered in this book, given sorting (one parameter at a time) in both the herbivore and producer communities. For simplicity, in generating the results summarized in this table, we assumed that water was not limiting (i.e., $W >> K_W$). The table also describes the impact of species sorting in the producer community in the absence of herbivores. One reads the table as follows: If a fat arrow points up (or down), this implies that species diversity (scaling the phenotypic variation available for sorting) for most parameters in that trophic level has a monotonic effect on a given ecosystem attribute (the rows); parameters that deviate from these overall trends are listed in lines below the arrows. If arrows point in both directions, effects may go in either direction, depending on exact parameter values. If an entry says "no effect," that means that for most parameters, diversity did not influence that aspect of ecosystem functioning. The table indicates qualitatively the range of potential effects of diversity on ecosystem characteristics, given bilevel trophic interactions.

In the absence of herbivores, the ecosystem properties of autotroph biomass, NPP, carbon storage, nitrogen mineralization, and evapotranspiration all tend to increase with species sorting at the producer trophic level. There are exceptions to this rule of thumb, however, when species differ in their basic density-independent rates of mortality. For instance, a decrease in death rates can increase biomass sufficiently to reduce NPP because of increased light competition.

In the presence of a single species of herbivore, many ecosystem processes also increase due to sorting among pro-

TABLE 11.1. Summary of Changes in Ecosystem Processes Given Sorting in Herbivore and Producer Communities

	System Open at Bottom			System Open at Top
	Producer Sorting		**Herbivore Sorting**	**Producer Sorting**
	No Herbivore	With Herbivore		
B^*	$\Uparrow$	no effect	$\Downarrow$	no effect
NPP	$\Uparrow$ $\mu \pm^a$	$\Uparrow$ $\mu\, 0$	$\Uparrow$ or $\Downarrow^d$	no effect $\mu \downarrow$
CS	$\Uparrow$	$\Uparrow$ $\mu \pm^b$	$\Uparrow$ or $\Downarrow^e$	no effect $\mu \downarrow$
$N_{\min}$	$\Uparrow$	$\Uparrow$ $\mu \pm^c$	$\Uparrow$ or $\Downarrow^f$	no effect $\mu \downarrow$
ET	$\Uparrow$	$\Uparrow$ $\mu, \gamma\, 0$	$\Downarrow$	no effect $\mu \downarrow, \gamma \downarrow$
H^* or H prod.	$\Uparrow$ $\mu \downarrow$	$\Uparrow$	$\Uparrow$ or $\Downarrow^g$	no effect

In this table, producer parameters are r, K_L, K_N, γ, and μ. Herbivore parameters are a', h, m, and c. The attack rate a' and handling time h can also be considered producer parameters, and producer sorting by varying these parameters has the opposite effect on the ecosystem attributes as does herbivore sorting. Therefore, conditions d through g below also apply to producer sorting on a' or h (second column), but with the inequality in the opposite direction. Producer sorting by a' or h increases all the ecosystem attributes except herbivore density and herbivore productivity (for which they have no effect) for the system open at the top (last column).

$^a\uparrow$ iff $r' < (\gamma + \mu)^2 K_L / \gamma$

$^b\uparrow$ iff $\dfrac{c}{m} > (1 - f)\left(\dfrac{1}{\lambda_s} - \dfrac{1}{\lambda_f}\right)$

$^c\uparrow$ with sorting by μ iff $\nu_h > \nu_p$

$^d\uparrow$ iff $B^* > \sqrt{r' K_L / \gamma} - K_L$, $r' = rD/(D + K_N \zeta)$, $B^* = m/[a(c\text{-}mh)]$

$^e\uparrow$ iff $B^* > \sqrt{\dfrac{r' K_L(1 + m / \lambda_f c)}{(1 + m / \lambda_f c)(\gamma + \mu) - m(1 + \mu f / \lambda_f + \mu(1 - f) / \lambda_s)\, c}} - K_L$ (except for c, m)

$^f\uparrow$ iff $B^* > \sqrt{\dfrac{r' K_L}{\gamma + \mu\, [1 - (\nu_p / \nu_h)]}} - K_L$

^{g}Both $\uparrow$ with sorting by a', c, h; H prod. $\uparrow$ with sorting by m iff $B^* > \sqrt{\dfrac{r' K_L}{\gamma + \mu}} - K_L$; $H^* \uparrow$ with sorting by m iff $B^* > \sqrt{\dfrac{r'(K_L ah - 1)}{ah(\gamma + \mu)}} - K_L$

TABLE 11.1. *Continued*

If the limit on B^* in condition e or f is complex (a negative value under the square root), then the herbivore sorting always decreases the ecosystem attribute, while for condition g, a complex value indicates that H^* always increases with herbivore sorting by m.

For herbivore sorting by changing c or m, in the system open at the bottom, the condition for increase in Carbon Storage is:

$$\left[(\gamma + \mu)\left(h + \frac{1}{\lambda_f} \right) - \left(1 + \frac{\mu(1 - f)}{\lambda_s} + \frac{\mu f}{\lambda_f} \right) \right]$$

$$(B^* + K_L)^2 > r'\left[K_L\left(h + \frac{1}{\lambda_f} \right) - \frac{1}{a} \right]$$

ducers. However, producer biomass does not increase, unless species vary in a' (the per capita rate of consumption by herbivores), in which case B^* increases as more-resistant producers replace less-resistant producers. Variation in this parameter can also potentially lead to declines in ecosystem attributes. As before, countervailing trends can arise if species vary in their direct mortality rates. Overall, adding herbivory weakens the impact of sorting at the producer level on ecosystem functioning.

Given a species pool of herbivores, and a single resident producer species, sorting favors the herbivore that can persist at the lowest plant biomass. Herbivore sorting tends to reduce evapotranspiration. Herbivore sorting can either increase—or decrease—NPP, nitrogen mineralization, and carbon storage. Overall, sorting at the herbivore trophic level tends to depress ecosystem attributes, opposite to the trends expected given sorting at the producer level.

Case II: Ecosystem Closed at Bottom, Open at Top

Now consider the same system, but with herbivore immigration and emigration at the top, and nutrient closure at the bottom (which we ensure by setting deposition and leaching both to zero). We consider only species sorting at the producer level, with a single herbivore species (but see Discussion). Table 11.1 describes the impact of such sorting on ecosystem attributes. Plant biomass is now unaffected by species sorting, unless species differ in their vulnerability to herbivory. A decrease in direct mortality implies a decline in NPP, carbon storage, nitrogen mineralization, and evapotranspiration. Decreases in herbivory increase all these ecosystem attributes. Interspecific variation in many of the other parameters describing resource uptake, respiration, and maximal growth has no effect on these ecosystem traits (sorting by γ does reduce evapotranspiration). The nature of coupling between the local system and the external environment thus qualitatively influences the mapping of species sorting at the community level on ecosystem functioning.

In short, the expanded basic ecosystem model reveals that the relationship of biodiversity to ecosystem functioning depends on (1) the presence or absence of trophic interactions; (2) the level at which biodiversity enters the system (viz., the contrasting effects of sorting at producer vs. herbivore levels), and (3) the precise nature of system openness.

DISCUSSION

We caution that these results all involve species sorting from regional pools where species differ along just a single dimension (represented by values of one parameter). More broadly, in natural systems one might expect coupling among parameters because of trade-offs (e.g., between resource assimilation and escape from herbivory). Incorporating trade-offs could well change the impact of sorting rules on ecosystem functioning. Moreover, and very importantly, trade-offs can permit local coexistence. For instance, if herbivores differentially attack a superior competitor among

256

the producers and ignore inferior competitors, inferior resource competitors can persist. In the limit of very effective herbivory, system properties should converge on those determined by the inferior competitor alone. We have also not considered sorting that occurs simultaneously at producer and consumer levels. Diversity at different trophic levels is expected to be mutually interdependent (Siemann et al. 1998, Knops et al. 2000), and a deeper understanding of ecosystem impacts of biodiversity surely requires an articulation of this interdependency.

The above protocol assumed that species sorting from competitive exclusion occurred on timescales that were short, relative to the rate of colonization from the regional species pool. A very simple way to permit local coexistence is for recurrent immigration from the species pool to occur at rates sufficient to offset local extinctions (Levin 1974, Holt 1993). Even if a single species is expected to dominate in a closed community, high species richness may be observed in an open community due to immigration of locally inferior species, comprising "sink" populations (Loreau and Mouquet 1999).

Permitting recurrent immigration (and emigration) has two distinct effects on ecosystem processes. First, it adds inputs and outputs to ecosystem compartments. We noted above that spatial linkages can qualitatively alter the expected relationship of biodiversity to ecosystem processes (see table 11.1). Second, immigration can counteract the directional effects of sorting via systematic changes in system parameters. Rather than discuss models with recurrent immigration in detail, we here summarize conclusions that will be presented more fully elsewhere (Loreau and Holt, in prep.).

Assume that there is no herbivory, with a local competitive dominance hierarchy in competition for a limiting resource, combined with immigration-extinction dynamics. The nutrient pool is closed, except via fluxes in the producers. If the producer species differ only in resource up-take rates, then there is *no* effect of diversity on primary production. An increase in producer diversity lowers the av-erage uptake rate, per producer (across individuals, in all species), but this is precisely compensated by an increase in resource availability. If instead, producer species differ only in their density-independent death rates, then production will be *higher* in more diverse systems. The reason is that if species with high death rates are maintained via immigra-tion, then more diverse systems have higher individual turn-over rates, therefore more rapid recycling of nutrients to the free resource pool. Finally, if species differ only in their rates of emigration, production is *lower* in more diverse sys-tems. Emigration is akin to mortality in determining local dominance, but also permits additional channels of nutrient loss from the system. More diverse systems can thus have a higher rate of resource drain from the local nutrient pool, ultimately lowering production.

The models we have considered assume very simple type I and II functional responses. Future work should include more realistic renditions of herbivore feeding behavior (Farnsworth and Illius 1998; Ginnett and Demment 1995; Schmitz, Beckerman, and Litman 1997). Over evolutionary timescales, plant-herbivore interactions can also strongly in-fluence detrital pathways for nutrients. For instance, Stein-berg, Estes, and Winter (1995) argue that the presence of the sea otter in the north Pacific, but not in the south Pa-cific, has led to systematic differences in the fraction of pro-duction devoted by algae to defensive compounds. They suggest that because sea otters were very effective at limiting invertebrate herbivores such as sea urchins prior to human impacts (Estes and Duggins 1995), over evolutionary timescales there has been much more intense herbivory in the southern hemisphere than in the northern hemisphere. Detritus produced by algae in the southern hemisphere may thus be less readily decomposed and support more attenuated detrital-based food chains. An evolutionary perspective may be necessary to explain some variation among studies on the observed impact of trophic interactions on ecosystem processes. For instance, Milchunas and Lauenroth (1993; see also Lennartson et al. 1997) suggestively found in a global survey of vertebrate herbivore exclosure experiments that the impact of grazers on primary production varied with the length of shared evolutionary history between producers and herbivores. Other topics needing examination in future work include impacts of higher trophic levels (e.g., trophic cascades) on ecosystem functioning, and detrital-based food webs.

CONCLUSIONS

As noted in the Introduction, systems vary tremendously in the detailed structure of food webs. Such variation is likely to be key to understanding the influence of biodiversity on ecosystem functioning. However, we suspect that the following insights we have drawn from the basic ecosystem model augmented with herbivory will apply much more broadly:

1. *Simple models may suffice to predict effects of species sorting on ecosystem functioning.* Many of the trends arising from species sorting in the "basic" but rather complex ecosystem model of the book also seem to appear in much simpler models (Holt in press). For instance, almost any model of competitive interactions among herbivores predicts that the dominant herbivore will be the one that can persist at the lowest abundance of producers. This will often imply overexploitation, and a lower abundance of the dominant herbivore. This in turn implies that species sorting at the herbivore level may initially increase total system biomass, but that eventually total biomass will surely decline as very effective consumers come to dominate the system. Detailed ecosystem models are undoubtedly needed to generate precise

predictions, but many key qualitative features may be captured and more clearly understood by analyzing related, simpler models.

2. *General trends in the effect of diversity on ecosystem functioning are present, but may differ among functional groups in the same ecosystem.* An important concern is the level at which sorting occurs versus the level at which the ecosystem effect is measured. For instance, in the system closed except for nutrient inputs, in most cases species sorting at the producer level increased ecosystem functioning. Species sorting at the herbivore level, by contrast, overall tended to reduce ecosystem functioning. There thus may be countervailing effects of diversity on ecosystem processes at different trophic levels.

3. *Effects of species sorting depend on the mechanistic parameter determining dominance.* Species sorting does not have an inevitable effect on ecosystem functioning. Different systems may show different relationships between diversity and ecosystem processes, if they differ in the axes along which competitive dominance is expressed. For example, table 11.1 shows that if a herbivore is present, and producers differ only in their basic mortality rate (with the producer with the lowest mortality rate being the best competitor), sorting toward lower mortality does not influence net primary production. However, if the producers instead differ in their basic assimilation rate of the resource, sorting toward more efficient producers increases net primary production.

4. *The presence of higher trophic levels influences the effect of species sorting on ecosystem functioning.* Given herbivory, the ability to withstand herbivory becomes an important component of plant competition. In some cases, sorting among producers toward lower herbivory rates can reduce ecosystem functioning.

5. *Spatial openness influences the mapping of species sorting onto ecosystem functioning.* The effects of diversity on system functioning (via species sorting) can be influenced by changing which compartment of the ecosystem is coupled

via flows to the external environment (see table 11.1). If species are maintained in local communities by immigration, then diversity in the regional species pool may have disparate effects on local ecosystem processes, depending on the parameter determining local competitive dominance.

6. *Species redundancy is a property not of species, but of species in particular systems.* Species are "redundant" with respect to certain system attributes, if their replacement does not change that attribute (Lawton and Brown 1993). In comparing the range of models we have considered, whether or not variation in a particular parameter of producer species leads to redundancy depends on the entire structure of the system (e.g., the presence or absence of herbivores, or the pattern of system openness). Two species may be functionally redundant in one setting, yet be nonredundant in another. As an example from table 11.1, if producer species in a system open at the bottom differ in basic mortality, and there is no herbivory, species sorting at the producer level increases biomass, nitrogen mineralization, and evapotranspiration. These species clearly are not redundant, because species substitution leads to a change in ecosystem functioning. However, if a herbivore is present, then species sorting for this same parameter has no effect on plant biomass or evapotranspiration. These species are thus redundant, given herbivory. More radically, if one compares systems open to spatial fluxes only at the bottom resource level with systems open only via herbivore fluxes, many species traits that are nonredundant in the former are redundant in the latter (in that species sorting does not alter ecosystem functioning).

Our most basic, qualitative conclusion is that in considering the impact of biodiversity on ecosystem functioning, a diversity of effects should be expected, depending on the presence and importance of trophic interactions, and the nature of system openness. This expectation of a diversity of outcomes is consistent with a recent review of empirical studies by Schläpfer and Schmid (1999), who reported considerable diversity among studies in the ecosystem effects of biodiversity. We end by conjecturing that some of this manifest heterogeneity reflects the varying importance of trophic interactions and system openness among ecological systems.

ACKNOWLEDGMENTS

RDH thanks the National Science Foundation for financial support, and ML thanks CNRS GDR DIV-ECO. RDH also thanks the Center for Population Biology, Imperial College at Silwood Park, where some of this work was carried out. We both thank NCEAS for its hospitality, and the organizers of this volume for their invitation, and their patience. Finally, we thank Michael Barfield for assistance in preparing table 11.1.

PART 3

Applications and Future Directions

CHAPTER TWELVE

Linking Soil Microbial Communities and Ecosystem Functioning

Teri C. Balser, Ann P. Kinzig, and Mary K. Firestone

INTRODUCTION

Microorganisms decompose organic matter and transform mineral nutrients in aquatic and terrestrial ecosystems. Despite the centrality of microbes, scientists often study ecosystem functioning without explicit reference to the microbial populations carrying out soil processes. The potential for rapid microbial growth, and the high degree of diversity and genetic exchange in microbial systems, has led to the often-held assumption that microbial catalysts do not limit the processes involved in ecosystem nutrient transfer and transformation (Meyer 1993). The utility of first-order kinetics in predicting process rates appears to confirm this assumption (Andrén, Brussaard, and Clarholme 1999). Recent advances in the techniques available to study microbial communities have, however, led to a growing realization that microbial community composition can play a critical role in ecosystem functioning (Schimel 1995; Schimel and Gulledge 1998; Cavigelli and Robertson 2000). First-order kinetic models may work well at equilibrium, but their utility in capturing process rates may be compromised as systems stray from equilibrium. In order to predict the transient response of changing ecosystems, and ultimately the process rates under new equilibrium conditions, we may need to understand and quantify the relationship between microbial community response and changes in soil processes such as decomposition and nitrogen mineralization.

In this chapter we explore when, how, and where microbial community composition affects ecosystem functioning. We begin with a short discussion of the challenges associated with linking microbial communities and ecosystem functioning. In light of these challenges and results, we review past analyses of the role of microbial community composition in ecosystem functioning, and present a conceptual model for advancing this field. Finally, we offer a blueprint for future research needs.

In much of our conceptual development and in most of our examples we discuss terrestrial systems and bacterial rather than fungal physiological ecology. Our conceptual model is, however, applicable to many situations and ecosystems. In part, the purpose of this discussion of the role of microorganisms in ecosystem functioning is to spark debate and new research in aquatic as well as terrestrial systems.

CHALLENGES IN LINKING MICROBIAL COMMUNITIES AND ECOSYSTEM FUNCTIONING

In 1993 an entire volume was published on the role of biodiversity in ecosystem function. At the end of 23 chapters the editors state: "From the present volume we can only conclude that soil biology is an area in which the concepts of ecosystem function and population biology have not been tested but which has a major importance for ecosystem integrity" (Schulze and Mooney 1993).

Our knowledge on these topics is still extremely limited. We do not yet have a firm understanding of how to characterize microbial genetic, physiological, or taxonomic diversity, much less an understanding of how to relate this diversity to controls on ecosystem processes and functioning. Even the common ecosystem model employed in this book—which proved extremely useful in extending the empirical findings of diversity–functioning relationships in plant communities—has serious limitations when it comes to examining the impacts of microbial communities on ecosystem functioning, in large part because there is only an implicit (decomposition and mineralization occur) rather than explicit microbial pool in the model. In the appendix to this chapter we examine what we can of the potential role of microbial community composition in influencing ecosystem functioning through use of the common ecosystem model, and highlight some of the challenges related to acquiring the additional information required to make a more sophisticated analysis of microbial community–functioning relationships.

Application of Macroscale Diversity Theory to Microorganisms

Microorganisms, with their predominantly asexual reproduction and ability to exchange DNA across 'species' boundaries, violate several of the starting assumptions of macroscale diversity theory. As applied to macroscale organisms, measures of genetic and taxonomic diversity traditionally rely on information about the number of species present (species richness) and the relative abundance of each species (species evenness) (Duelli 1997; Griffiths, Ritz, and Wheatley 1997). But we cannot easily and accurately determine the number and relative abundance of microorganisms in the soil, nor characterize species' boundaries in these systems (Stahl 1995; Dykhuizen 1998). First, the definition of a bacterial species is problematic; at present we define a bacterial species as composed of organisms that have 97% similarity in their 16s ribosomal RNA sequences. This measure is very coarse; Staley (1999) likens it to defining humans and chimpanzees as the same species, simply because our RNA is more than 97% similar. Second, even if we accept the definition of species as it currently stands, quantifying the richness of bacterial and fungal 'species' in the soil is prohibitive. There are over 5,000 microbial species in a gram of even the most common soil, and novel gene sequences are being discovered as new soils are studied (Torsvik et al. 1990; Meyer 1993; Borneman and Triplett 1997; Tiedje et al. 1999). We cannot easily get a handle on the number of species in a typical soil ecosystem. Finally, using gene-based assays (arguably the best current approach to microbial diversity), we are unable to easily quantify the evenness of microbial species. Quantitative analysis of DNA in soil is difficult (Suzuki and Giovannoni 1996; Polz and Cavanaugh 1998). Thus, as with species richness, it is theoretically possible to measure evenness, but it is as yet difficult. Thus macroscale concepts of diversity are problematic, if not irrelevant, for microbial communities.

Microbial Ecology Contribution to the Study
of Ecosystem Functioning

Many microbiological studies utilize molecular methods to catalog community diversity or to assess community response to various stresses and disturbances (Burkhardt et al. 1993; Giller et al. 1997; Knight, McGrath, and Chaudri 1997; Meyer et al. 1998; Wünsche, Brügemann, and Babel 1995; Tiedje et al. 1999; Andrén and Balandreau 1999). These studies document a stunning amount of microbial genetic diversity in a wide range of ecosystems (Nusslein and Tiedje 1998, 1999; Dykhuizen 1998; Torsvik et al. 1990; Borneman et al. 1996; Borneman and Triplett 1997; Dunbar et al. 1999). Nusslein and Tiedje (1998) find that the diversity of microbial DNA in an extremely young tropical soil (300 years old) is equal to that of soils tens of thousands of years older, with fully developed plant communities. This seems to suggest that soils can attain high microbial genetic diversity rapidly, and, thus, that microbial genetic diversity may not be a major factor limiting ecosystem functioning.

These approaches, however, tend to concentrate on ob-

268

taining measures of microbial community diversity without including other characteristics of the microbial community. In particular, there has been little emphasis on the need to quantify the relationship between microbial community characteristics and soil processes. The measures of diversity being formulated for microbial communities usually fail to describe their metabolic talents, their ability to respond to environmental determinants, their interactions with one another, or their spatial distributions throughout the soil matrix. This omission largely reflects the difficulty of relating the results from current molecular methods to the functions of the populations being examined. Microbial genetic diversity will continue to be an extremely interesting area of research, but establishing quantitative and causal links between microbial communities and ecosystem functioning may require new concepts and approaches.

Ecosystem Science and Microbial Ecology

In direct contrast to microbial ecology, ecosystem research has generally focused on processes in soil such as decomposition, nitrogen cycling, or trace gas production without explicitly considering the populations of microorganisms engaged in these processes. Ecosystem scientists are aware that microorganisms play a critical role in ecosystem functioning. Yet, because microorganisms are everywhere on earth, and because the processes they carry out are roughly the same everywhere, ecosystem scientists have generally assumed that microbial communities are functionally and compositionally interchangeable (Meyer 1993). In light of recent discoveries of seemingly limitless diversity of microbes it is easy to assume that microbial potential will not limit ecosystem functioning. However, as we will discuss, this assumption may not hold for many ecosystems and/or future global change scenarios.

In the general ecosystem model presented in this book, a

given alteration in plant diversity is coupled with measurable or predictable changes in ecosystem functioning, such as productivity or nitrogen cycling. The assumption is that changes in macroscale diversity are both likely in the face of current global changes and meaningful with respect to ecosystem functioning. Microbial diversity differs fundamentally from plant diversity in this respect. Except in the case of extreme environments or catastrophic events, significant reductions in microbial diversity are unlikely. This does not imply that we need not concern ourselves with the study of microbial community composition as related to ecosystem functioning, rather it highlights that the primary challenge in linking microbial diversity and ecosystem functioning may be to avoid thinking of microorganisms in the same way that we think of the macroscale community. We suggest that working with the broader concept of microbial community composition may be more useful than microbial genetic or taxonomic diversity.

Recent work has demonstrated that microbial community composition controls soil process rates independent of environment (temperature and soil water) (Balser 2000). Measures of the phenotypic composition of a microbial community may provide a useful link between microbial communities and ecosystem functioning. Stark and Firestone (1996) found that differences in the temperature adaptation of nitrifier populations could determine nitrification rates in the California annual grassland. Cavigelli and Robertson (2000) found that denitrifier populations in never-tilled successional and conventionally tilled fields in Michigan had different degrees of sensitivity to pH and soil water and thus constrained nitrous oxide flux. Similarly, Gulledge and Schimel (1998) found that differential sensitivity to water stress and adaptation to local moisture regimes of methane oxidizer populations in Alaskan soils constrained methane consumption.

LINKING MICROBIAL COMMUNITY COMPOSITION
AND ECOSYSTEM FUNCTIONING:
A REVIEW OF CONCEPTS AND MODELS

Broad versus Narrow Processes

Schimel (1995) proposed that the role of microbial community composition in ecosystem functioning is related to the characteristics of the populations carrying out the processes. He proposed that ecosystem processes are based on broad or narrow physiological groups of microbes. "Broad" processes would include soil respiration, mineralization of simple carbon compounds, and nitrogen immobilization, all processes requiring metabolic capacities possessed by many seemingly redundant populations. Most microorganisms utilize simple carbon substrates and all require nitrogen for biosyntheses. Schimel predicts that these broad processes will not vary much within or between systems. In contrast, "narrow" processes are those carried out by restricted groups of organisms. Examples are nitrification, trace gas production and consumption, and lignin degradation. For these types of processes a change in abundance or characteristics of populations with the given metabolic capability could have an observable impact on process at the ecosystem scale.

This work by Schimel (1995) is among the first to propose a framework linking microbial community composition to ecosystem functioning, and it works reasonably well in explaining the results from studies to date. Studies by Stark and Firestone (1996), Cavigelli and Robertson (2000), Gulledge and Schimel (1998), Dendooven, Pemberton, and Anderson (1996), and Balser (2000) all show community-specific characteristics of narrow groups impacting process rates at the ecosystem scale. However, the conceptual model of broad and narrow processes primarily addresses equilib-

271

rium conditions. Furthermore, definitions of narrow and broad processes can become difficult. Is cellulose degradation, easily carried out by many genera, a broad process? Or is it narrow, because it requires the action of a specific consortia of degraders? Are narrow processes phylogenetically defined? Or, like denitrification, do they span phylogenetic clades? Below we propose a conceptual framework based on simple principles of microbial physiological ecology and adaptation to their environment.

Application of Physiological Ecology

In typical ecosystem models, microorganisms are catalysts with unchanging properties (Parton et al. 1987). In more detailed models, the quantity of catalyst (biomass) is allowed to change, but its properties remain constant over time (McGill et al. 1981; Kinzig 1994). Microbes are not inorganic catalysts, however—they are dynamic biological organisms whose physiology and interactions with the environment are subject to many of the same constraints as those of higher organisms. Thus microbes employ adaptive strategies and trade-offs to survive. In response to environmental change, both their biomass and community composition will change.

Microbes, for instance, are known to have growth optima (figure 12.1; Brock and Madigan 1991; Atlas and Bartha 1993), and when they are forced to operate near the limits of their range, they are less efficient (less biomass is produced per unit carbon utilized) (Balser 2000; Anderson 1994; Anderson and Domsch 1993; Wardle and Ghani 1995). When the soil environment changes, the composition of the microbial community will shift as organisms better suited to the new environmental conditions grow and compete, or as existing organisms undergo evolutionary adaptation (figure 12.1).

In a reciprocal soil transplant between two climates (Balser 2000), Balser found that the temperature tolerance of

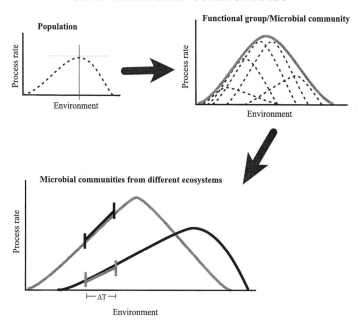

FIGURE 12.1. Hierarchy of growth or process response curves: population to ecosystems. (A) Populations have environmental limits and optima. (B) Many populations, each with a characteristic response curve, make up a microbial community. The curve resulting from each individual response is the whole-community response curve, indicating whole-community physiological flexibility. (C) Microbial communities from different ecosystems have characteristic whole-community curves. Microbial communities differ in sensitivity to environmental change and disturbance, and thus process rates differ after change or disturbance.

transplanted grassland and mixed-conifer forest soil microbial communities changed from that characteristic of their native climate toward that of the indigenous microbial community in the new climate (figure 12.2). In a warmer climate, the conifer forest soil community became able to survive and grow at higher temperatures, while the grassland soil, which was moved to a cooler climate, showed a decreased maximum temperature tolerance.

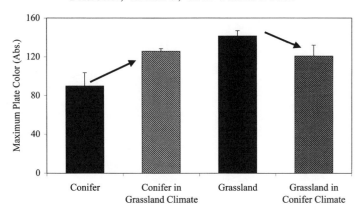

FIGURE 12.2. Change in temperature tolerance after two-year transplant: change in BiOLOG™ plate color development at 35°C. Error bars indicate standard error of the mean ($n = 5$). Conifer soil in the grassland climate grows better at 35°C than conifer in place (t-test, $p < 0.05$).

While these changes in the soil microbial community are predicted by basic ecological understanding, we actually know little about the time frame over which microbial community shifts occur and whether such changes in the microbial community impact the rates of processes or in fact constrain ecosystem functioning. It has long been assumed that the timescale of microbial response is short compared to the environmental change itself. In the soil transplant study, however, we found that microbial community composition was surprisingly resistant to change, even though the temperature response of the microbial community changed (figure 12.3). After one and two years of incubation in a new climate, the conifer soil in a warmer climate remained indistinguishable from the conifer incubated in its native climate (figure 12.3A and B). Grassland soil showed a slight shift in its fatty acid principal component signature. This was most likely due to unintentional inoculation by fungal spores in the conifer site (data not shown). To model or predict microbial mediation of ecosystem processes in the

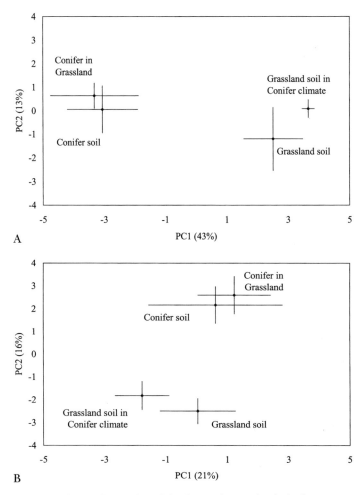

FIGURE 12.3. Ordinate plot of the first and second principal components from microbial community fatty acid profiles (A) 12 months and (B) 27 months after reciprocal transplant between conifer and grassland ecosystems in the southern Sierra Nevada of California. Numbers in parentheses indicate the amount of variance in the data accounted for by each component. Points are the mean of five replicates. Taxonomically similar communities plot close together, different communities are spread apart on an ordinate plot.

face of global changes, a more robust conceptual and mechanistic framework for the timescale of microbial response may be required.

Microbial Strategies: Physiological Constraints and Trade-Offs

The physiological "talents" of a microbial community (the community's phenotypic composition) are genotypically determined through the adaptation of member populations to selective pressure. No one organism, however, can respond to all environmental contingencies. An organism with broad physiological flexibility will be at an advantage where the environment changes rapidly and often. If, however, the environment rarely changes, then there is a disadvantage to having a broad functional range (i.e., having more genetic material to maintain) (Slater and Godwin 1980; Halverson, Jones, and Firestone 2000). Organisms from stable environments might be expected to have narrower ranges of physiological flexibility and therefore a decreased ability to respond to environmental changes. The assemblage of microbial populations in a typical ecosystem would likely contain a mix of generalists and specialists, resulting in a characteristic overall community sensitivity to environmental conditions (figure 12.1).

In the transplant study introduced above we found that soil microbial communities from grassland and mixed-conifer ecosystems had different sensitivities to a change in climatic conditions. The California annual grassland experiences a much broader range in temperature and water annually than does the Sierran mid-elevation mixed-conifer forest (Trumbore, Chadwick, and Amundson 1996). In our study, the conifer forest microbial community was more sensitive to a change in climate than was the grassland soil microbial community (Balser 2000).

The interaction of physiological plasticity and environmental determinants leads to the optimization of microbial growth and activity under a given set of environmental con-

276

ditions. Below, we briefly introduce some concepts of microbial physiological ecology that are important in constructing a dynamic conceptual model of microbial communities and ecosystem functioning.

MODULATOR STRESS

Modulators are environmental factors not necessary for microbial growth. They include water potential, temperature, pH, salinity, sodicity, and soil matric potential. Microorganisms exhibit biological homeostasis in response to changes in modulators—maintaining, for instance, nearly constant internal pH, or altering their internal solute potential in response to a change in soil salinity. It requires energy, however, to adjust cell integrity when a modulator strays from the levels at which growth is optimal. This response is common to nearly all soil microorganisms. Thus, the stress resulting from a change in the characteristics of a modulator should affect the whole microbial community.

RESOURCE AVAILABILITY

In contrast to modulators, resources are those things required by organisms for growth (e.g., carbon, nitrogen). In the constantly changing soil environment an organism will respond rapidly to either resource excess or limitation. Many organisms respond to resource excess by forming certain "storage" compounds (Dawes 1989), which may serve as energy or nutrient reservoir against times when the stored resource becomes limiting (Chapin et al. 1987). Response to resource *limitation* is more complicated. The general response strategy is designed to allow an organism to take up and metabolize a limiting resource at the fastest rate possible, and to produce a maximum amount of cell material with a minimum of the limiting factor (Harder and Dijkhuizen 1983).

Resource excess or limitation will ultimately affect microbial community composition. What is a resource for one mi-

277

crobial population may not be for another. Thus resource fluctuations affect the microbial community through stimulation or depression of specific populations. For example, nitrate added via atmospheric deposition will allow organisms capable of utilizing that nitrate in denitrification to increase numerically relative to organisms who do not utilize nitrate as a resource.

IMPORTANCE OF ADAPTATION TO NATIVE MODULATOR AND RESOURCE LEVELS

Adaptation of microbial communities to local equilibrium climate (modulators) and resource regimes conditions their response to perturbation. For instance, microorganisms in a cold climate are able to respond quickly to a change in temperature of a few degrees, as arctic ecosystems experience a very short growing season (Nadelhoffer et al. 1991; Panikov 1999). Populations in a community adapted to a more stable temperature regime, however, do not respond as quickly (Balser 2000; Raich and Schlesinger 1992). Similarly, Sugai and Schimel (1993) and Hunt et al. (1988) report the importance of microbial adaptation to carbon resource characteristics: litter that is placed in its native habitat is degraded more readily than litter in a habitat where the microbial community is unaccustomed to it. Thus, the response of a microbial community to altered modulator or resource regimes will depend on the initial characteristics of those regimes, and the microbial adaptations to them.

TIMELINE OF MICROBIAL RESPONSE: CONCEPTUAL MODEL OF MICROBIAL ROLE IN ECOSYSTEM FUNCTIONING

Microbial Response: Four Phases

In Balser's study of transplanted soil, soil process rates over the two-year study period were more closely related to microbial community composition than to environmental

278

TABLE 12.1. Effect of Microbial Community versus Environment on Soil Processes

Y variable	X variable	R^2	$p > F$	Independent Effect of X Variable, as a % of R^2
N_2O	T	0.476	<.0001	13.83
	Water			13.16
	PC1B*			1.69
	PC1F*			71.32
Nitrification	T	0.141	0.0394	5.20
Potential	Water			0.26
	PC1B			5.29
	PC1F			89.25
CO_2	T	0.565	<.0001	64.74
	Water			1.35
	PC1B			25.45
	PC1F			8.45
Net	T	0.246	0.0009	14.61
Mineralization	Water			32.75
	PC1B			12.53
	PC1F			40.12
Gross NH_4^+	T	0.188	0.0081	14.78
Mineralization	Water			13.76
	PC1B			4.38
	PC1F			67.08

Notes: We determined the independent effect of each of four X variables on a given soil process variable (Y variables) by partitioning the variance in the R^2 term among the X variables (determined the "part correlation"; Selvin 1995). This gives an estimate of the importance of each X variable in the relationship, *independent* of the other X variables. We report the results as a percentage of the overall R^2 for the regression.

* PC1B and PC1F are the first principal components from BiOLOG and PLFA assays of the soil microbial community, respectively. We use the first principal component as a summary variable that best represents the functional (BiOLOG) or taxonomic (PLFA) community profile (Selvin 1995; Balser 2000).

factors (table 12.1). Using the first principal component from phospholipid fatty acid analyses and BiOLOG™ sub-strate utilization as univariate canonical summary variables for the microbial communities (Selvin, 1995), we (Balser and Firestone) found that for N_2O production, nitrification

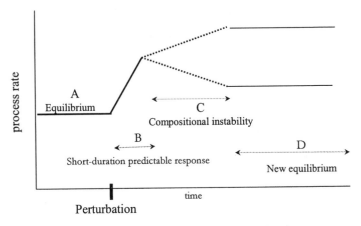

FIGURE 12.4. Response timeline. A: Steady state (functional groups catalyze processes at rates resulting from optimization to ambient environmental conditions); B: Predictable response (after perturbation, process rates will change rapidly—potentially predictable by chemical kinetics and diffusion); C: Community compositional change (populations acclimate and adapt; dominant members of the microbial community may change. Eventually, a new "steady-state" community will be reached); D: New steady state (successional microbial community "optimized" for new conditions). The amount of time it takes to move from B to D is not known and likely varies among ecosystems.

potential, net N-mineralization, and gross NH_4^+ mineralization, microbial community variables explained more of the process variability than did either soil temperature or water content. This has implications for the way that we model ecosystem dynamics: most current models do not allow for microbial community evolution or time lags in compositional or functional change. First-order rate constants are derived from microbial community processes under existing conditions, presumably at equilibrium. Yet under changing conditions the microbial community adapts in response to new selection pressure; this will likely result in a change in rate constants. We propose a timeline of microbial response to disturbance or change that has four phases (figure 12.4). During each phase microbial community composition may impact soil processes.

SOIL MICROBIAL COMMUNITIES

STARTING EQUILIBRIUM (A)

Process rates are determined by microbial community adaptation to local environment, including climatic regime (normal seasonal range) and carbon inputs (Balser 2000; Hunt et al. 1988; Clein and Schimel 1995; Schimel and Clein 1996). Thus a tropical forest microbial community has a different temperature response (e.g., Q_{10}) than a tundra community.

PERIOD OF PREDICTABLE RESPONSE (B)

For a short time after perturbation, effectively before community composition changes, process rates may be predictable based on the change in the physical environment and the rate constants derived for the microbial community at equilibrium. Thus, over a short time period, the effects of changing environment can be modeled or predicted from first principles, or from previous experience. For example, the effect of a change in temperature on microbial processes can be reasonably modeled as temperature effects on the velocity of a chemical reaction (hence the utility of Arrhenius equations) or from empirically based Q_{10} values (Paustian et al. 1997). Similarly, the effect of changing moisture can be quantitatively predicted by changes in substrate diffusion as a function of water film thickness (Stark and Firestone 1995; Holden 1995). Both of these approaches for predicting the effects of temperature and water are valid—for at least several hours. After an initial short period of time, however, acclimation and adaptation of the microbial species must be expected to override the simple, mathematically predictable rate responses.

This type of predictable rate response has been included in models of the effect of global climate change on terrestrial ecosystems. If models assume no change in microbial community composition in response to changing environment, then empirically derived rate constants are valid indefinitely. If, however, the microbial community changes in composition, or species adapt in response to change, then

281

constants used in simple predictive rate equations must change.

COMMUNITY COMPOSITIONAL CHANGE (C)

Populations affected by changing conditions will acclimate and adapt, ultimately to outcompete or be overwhelmed by other microbial populations. This is a period of compositional instability during which process rates may effectively be independent of substrate pool size or environmental drivers. Microbial adaptation may be accompanied by changes in net community kinetic constants, such as those predicting rate maxima, and rate dependence on substrate availability. During this period, then, many of the "constants" in first-order rate equations will change. Rates may become dependent on the number of microbes catalyzing the process and thus cease to be predictable by simple first-order kinetics derived from previous equilibrium conditions.

NEW EQUILIBRIUM (D)

Eventually, after exposure to different environmental conditions, microbial communities become adapted, and their processes reach a new equilibrium. Rate constants for processes catalyzed by the microbial community will now be characteristic of this adapted community. The composition of this new microbial community will be framed both by the characteristics of the ambient environment and the composition of the antecedent community.

Microbial Community Response to Modulator versus Resource Change

Previously we discussed the importance of resources versus modulators. In this section we discuss how microbial community response to a change in resource or modulator levels is related to our timescale model, and we try to put that response into the perspective of current global changes.

Our proposed timeline of response is generalized. It is likely that the response of microorganisms to change in their environment will differ for a change in modulator versus a change in resource availability. Prior adaptation of the microbial community to native resource and modulator levels will affect the impact of a change in microbial community composition on process rates.

A change in resource availability, such as nitrogen addition to an N-limited system, will have a different effect on microbial community composition than a change in environmental conditions (modulators). A change in resource availability will likely impact a specific subset of the microbial community, causing a comparatively large change in select populations. For example, microbial communities subjected to increased nitrate levels become dominated by populations capable of readily using nitrate, resulting in increased denitrification (Verchot, Franklin, and Gilliam 1998), or NO_3^- assimilation. In the presence of excess NH_4^+, ammonia oxidizer populations increase, producing larger nitrate pools. This increased importance in the microbial community of populations that are normally a lesser component can result in faster nitrogen cycling and increased NPP or N loss from the ecosystem. Because of its effect on specific populations and the rapid growth potential of microorganisms, we suggest that the timescale of microbial community response to a change in resource availability will be short relative to the time required for community response to a change in modulators.

In contrast, a change in modulators will affect the microbial community as a whole, rather than specific populations. Instead of a subset of the microbial community becoming dominant by outgrowing and outcompeting others for a resource, the microbial community will change more gradually as those organisms better able to tolerate the new conditions come to dominate the community, or adapt to the new environment. This is the type of response we saw in our two-year soil-transplant experiment. The whole community from either conifer or grassland soil in its native climate remained compositionally similar to the transplanted community, yet the temperature response curve gradually changed.

Relevance to the Timescale of Global Changes

Because microorganisms studied in the lab have generation times of hours, it is often assumed that their response will "keep pace" with global changes; that is, the microbial community will track environmental conditions or plant communities. In general, this may be true. Over long enough periods of time, microbial community composition will be under environmental control. Where there are ecosystems in equilibrium, there is an associated microbial community that is also in equilibrium. However, the microbial community does not respond instantaneously to environmental or plant community change. We present data that suggest that microbial community adaptation to native environmental and resource regimes can cause a time lag in response to changes in environment or resources. This time lag might play a critical role in microbial community–ecosystem relationships. Sudden changes in plant community composition or environment (such as land use and land cover changes) can create a disequilibrium between the microbial community and carbon and nitrogen cycling. The resulting fluctuations in decomposition nitrogen/nutrient availability might be enough to affect the trajectory of plant community succession. In Balser's (2000) transplant study, the rapid change in environmental conditions when soil cores were transplanted mimicked a change in land use or cover. The nitrogen cycle in affected conifer soil was altered for two years. The grassland soil, adapted to a more fluctuating environment, was less affected. Thus, these relatively "short-term" transients in microbial community response may be enough to fundamentally alter soil nutrient status, constraining the types of plants that can persist in the system and, ultimately, the functioning of the system.

CONCLUSIONS AND FUTURE RESEARCH NEEDS

The continuing debate over biodiversity–ecosystem functioning relationships, and the plant and microbial contributions to those relationships, highlights some of the difficulties inherent in trying to disentangle the influences of diversity on processes. The ecologists engaged in empirical demonstrations of diversity–process relationships have largely been concerned with examining the influence of variations in species number or community composition when all such assembled communities are subjected to otherwise similar climatic and initial edaphic conditions. That is, they manipulate diversity without manipulating environmental conditions. For microorganisms, manipulation of diversity is artificial and misleading. Microbial genetic diversity is extremely difficult to measure, and is of questionable relationship to process measures in soil.

The analyses presented in this chapter suggest that a new and different approach is needed for microbial communities—examining how given microbial communities respond to permanent or sudden environmental changes, how these responses influence ecosystem processes, and how or whether these responses and influences are related to initial microbial community composition or diversity. This is only minimally related to the analyses that are conducted concerning the influence of diversity on stability and resilience, which, after all, imagine a system perturbed from its initial conditions and examine the return to those conditions. The ecosystems of this century may be continually stressed, continually prevented from reaching any sort of equilibrium because of constantly changing climatic, resource, and edaphic conditions created by anthropogenic activities. It is under these "permanent" or long-term transition conditions that the influence of microbial community on ecosystem processes may be most keenly felt.

Why might that be? Microorganisms are commonly assumed to be ubiquitous—and able to respond quickly to environmental variations that are within the range of those commonly visited upon given systems. Under these conditions, microbial community composition and number would not constrain process rates—environmental conditions would influence those rates.

It is also possible, however, that the expected "rapid response" of the microbial community may fail when these communities experience novel conditions. The grassland–conifer soil transplant cited in our chapter demonstrates this point: the mixed-conifer soils, adapted to a narrow range of climatic conditions, still had microbial communities of conifer systems—and process rates differing from grassland systems—two years after being transplanted to grassland systems. We do not know how the time dynamics of microbial response might compare to the time dynamics of environmental change. Will microbial communities be able to "keep up"—and thus not exert a constraint on process rates—under gradual environmental change? Even if the microbial community equilibrates rapidly relative to the plant community, it is still entirely possible that the characteristics of that equilibrium microbial community will influence and ultimately determine the trajectory of the plant community. For instance, microbial controls over nitrogen cycling will influence the amount of nitrogen available to plants—and thus the kind of plants that can invade or persist in a system. The relevant question in this case, then, is whether initial differences in microbial community phenotypic diversity or composition influence the final equilibrium characteristics of the microbial community (under the new conditions), or whether the characteristics of this new community are determined solely by the final climatic or edaphic conditions. Will we always find similar microbial communities under similar conditions, thus indicating that the conditions themselves are the final controllers of ecosystem processes and functioning? Much of the work presented in this chapter suggests that microbial community composition might matter, that microbial communities do not always "keep up" with prevailing conditions. It may be the characteristics of the microbial communities themselves, as well as climatic or edaphic conditions, that determine eventual outcomes. Again, more questions than answers remain, but we would conclude on the basis of arguments in this chapter that the influence of microbial community composition on ecosystem processes should be examined not under constant environmental conditions but when environmental conditions are changing.

ACKNOWLEDGMENTS

Balser and Firestone would like to thank the Kearney Foundation for Soil Science Research for financial support, and Egbert Schwartz for valuable feedback and discussion.

APPENDIX

Linking Microbial Community Composition and Ecosystem Functioning: Incorporating Microbial Dynamics in the Common Ecosystem Model

The common ecosystem model used throughout the theory chapters in this monograph (see chapter 8) as structured does not allow for a direct analysis of the impacts of microbial community composition on the ecosystem processes of interest. Microorganisms are not explicitly included in the model—instead, their presence is only implicit, through the functions of decomposition and mineralization. Nor are microbial dynamics particularly complex in the common ecosystem model—decomposition rates scale linearly with substrate, and there is no explicit microbial immobilization of inorganic nitrogen (C:N ratios of the fast and slow carbon pools remain constant). In this section, we consider two additional cases of plant diversity–process relationships for the competition-colonization model that alter these assumptions about microbial roles in ecosystem processes. (See chapter 9 for a more complete description of the features of the competition-colonization model and the previous cases of diversity–functioning relationships considered.)

CASE A: NONLINEAR DECOMPOSITION DYNAMICS

In the basic ecosystem model, decomposition is proportional to carbon stocks for both the fast and slow pools, regardless of the quantity of carbon stocks. A more realistic model would include a saturation of the decomposition rate at high levels of soil carbon (this might occur if, for instance, microorganisms are initially able to increase in abundance due to the availability of carbon, but eventually reach some carrying capacity, due to other resource limitations, predation, etc.). In order to examine the effects of this phenomenon, we assume that the rate of decomposition for the slow carbon pool now goes as $C_s K_{th} \lambda_f / (C_s + K_{th})$ (as opposed to the previous expression of $C_s \lambda_f$). All other parameters are as in the Case 3 analysis in the competition-colonization section of chapter 9 (closed nitrogen system, inclusion of performance trade-offs).

CASE B: MICROBIAL IMMOBILIZATION

In the common ecosystem model, the C:N ratios of the fast and slow carbon pools never change—they are set at 15 and 225, respectively, in the competition-colonization analysis. In reality, the C:N ratios of various carbon pools in the soil system will change as microorganisms "gas off" excess carbon, immobilize available nitrogen, or release nitrogen upon death. To simulate these dynamics, we allowed the C:N ratio of the slow carbon pool to vary. In particular, we assumed that carbon availability controlled microbial population size, and that decomposition of the slow carbon pool would saturate when microorganisms reached their carrying capacity. Below that threshold, microorganisms would grow with additions of carbon to the slow pool, and would have to immobilize available inorganic nitrogen to do so (since the high C:N material in the slow carbon pool does not contain enough nitrogen to permit growth of low C:N microorganisms). When microbial populations exceeded their carrying capacity, population decline would occur and microorganisms would release inorganic nitrogen. In these simulations, therefore, less nitrogen was available to fuel plant growth relative to the previous simulations, since more nitrogen is bound up in microbial biomass.

RESULTS FROM CASES A AND B

In both of these cases we attempt to include a higher degree of microbial realism by allowing for dynamic microbial populations. The results are shown in table 12.2. The only relationships qualitatively affected (relative to the results reported in chapter 9) are the upper bounds of performance for total carbon and net primary production—the best performance now either remains relatively constant (total carbon) or decreases (NPP) with increasing plant diversity. The introduction of a saturating decomposition rate in the slow carbon pool affects the superior plant competitors more than the inferior ones, since the superior plant competitors are assumed to be "woodier" species and thus allocate more of their tissue to the slow carbon pool upon death. Thus, decomposition and N-mineralization are depressed for those plant types that have higher C:N ratios—and NPP and plant biomass are therefore depressed relative to the cases where decomposition rates do not saturate. Adding superior plant competitors to the best-performing monoculture, therefore,

Table 12.2. Relationship between Diversity and Ecosystem Process in the Various Cases Considered

Function	Feature	Case 1: Random Fecundities (chap. 9)	Case 2: Other Performance Trade-offs (chap. 9)	Case 3: Closed Nitrogen Cycle (chap. 9)	Case A: Saturating Decomposition Dynamics (this chap.)	Case B: Microbial Immobilization (this chap.)
Total C	Average performance	Increasing[a]	Increasing	Increasing	Increasing	Increasing
	Envelope (upper bound)	Flat[b]	**Increasing (weak)**[d]	Increasing	**Flat**	Flat
NPP	Average performance	Increasing[a] (weak)[c]	Increasing	Increasing	Increasing (weak)[c]	Increasing (weak)[c]
	Envelope (upper bound)	Declining	**Increasing (weak)**[d]	Increasing	**Declining (weak)**[d]	Declining

N-min	Average performance	Increasing[a]	**No relationship**	No relationship	No relationship	No relationship
	Envelope (upper bound)	Flat[b]	**Declining**	Declining	Declining	Declining
Evap	Average performance	Increasing (weak)[a,c]	Increasing (weak)[c]	NA[e]	No relationship	No relationship
	Envelope (upper bound)	Flat[b]	**Increasing (weak)[d]**	NA	Declining	Declining

Notes: Boldface indicates a qualitative change in either the average performance or the envelope under a change in assumptions.

[a] In these cases, there is actually a slight downturn in function at higher levels of diversity—that is, the average performance for the 9- and 10-species communities are actually lower than the average performance for the 8-species community. At lower levels of diversity, there is an increase of ecosystem process with diversity.

[b] The best performance is relatively flat over the range of 1- to 8-species communities, but declines with 9- and 10-species communities.

[c] The increase in process is labeled "weak" if the average performance of all monocultures is within 15% of the average performance of all 10-species communities.

[d] The decline or increase in best performance is labeled "weak" if the best performance of the 10-species community and monoculture are within 15% of each other.

[e] The water model was excluded in this analysis, and thus diversity-process relationships for evapotranspiration were not considered.

may decrease functioning, as the superior competitors are poor performers and reduce the abundances (and therefore performances) of the intermediate competitors.

Introduction of microbial immobilization, at least in the form considered here for Case B, does not change any of the qualitative features of the plant diversity–ecosystem functioning relationships beyond those found in Case A. Again, the introduction of such a dynamic affects the superior plant competitors more than the inferior ones, as the immobilization is assumed to occur in the slow carbon pool, to which the superior plant competitors contribute more of their biomass upon death.

Do these results mean that "microorganisms don't matter"? No definitive conclusions can be drawn from this cursory analysis. What is remarkable among this set of cases, and those considered in chapter 9, is the relatively robust forms of the plant diversity–functioning relationships, in spite of variations in assumptions. Thus, the "increase" in microbial diversity in these simulations—from one class of microbes to two—did little to affect the form of the plant diversity–functioning relationship. The functions themselves are affected by the change in assumptions—more carbon can be stored in the cases where decomposition saturates at high carbon levels, and less primary production can be maintained when more nitrogen is bound in microbial biomass. Thus, the ways in which microorganisms are behaving can influence functioning, but considering their diversity per se (as two "classes" of microbes rather than one) in these equilibrium analyses does not offer any new insights into diversity–functioning relationships.

We have not, however, considered possible relationships between plant diversity and microbial community composition. Different microbial communities may exploit a wider variety of carbon substrates, and thus elevate rates of mineralization and net primary production. If greater plant diversity begets greater microbial diversity, then decomposition rates may actually increase not only with increasing substrate but with increasing plant diversity as well. This effect can also be mimicked within the common ecosystem model, by assuming that decomposition rates—for a given level of substrate—increase with increasing plant diversity. If the increase in decomposition rates with plant diversity is great enough, then the decline in performance with diversity for net primary production and N-mineralization can, not surprisingly, be reversed (see table 12.2). This suggests a more detailed analysis is needed of the relationships between plant diversity and microbial community composition, and the implications for diversity–functioning relationships.

Nonetheless, these results suggest that a more explicit inclusion of microbial processes in our models may do little to influence or improve our understanding of diversity–functioning relationships under *equilibrium* conditions. We do not want to argue that microbial communities are not important participants in key ecosystem processes—it is self-evident that they are. The question before us, however, is whether microbial community composition per se regulates or controls ecosystem processes. In this chapter, in fact, we extended our analysis beyond the equilibrium conditions considered here, and offered some initial thoughts on the likely conditions under which microbial community composition might most matter.

CHAPTER THIRTEEN

How Relevant to Conservation Are Studies Linking Biodiversity and Ecosystem Functioning?

Sharon P. Lawler, Juan J. Armesto, and Peter Kareiva

> Suppose one could make out a good case for
> conserving the variety of nature on all three
> grounds—because it is a right relation between
> man and living things, because it gives
> opportunity for richer experience, and because
> it tends to promote ecological stability?
> —Charles Elton, 1958

INTRODUCTION

Research on biodiversity and ecosystem functioning has been embraced by some in the conservation community (e.g., Walker 1995; Edwards and Abivardi 1998) but viewed with skepticism by others (e.g., Soule 1996; Schwartz et al. 2000). These differing attitudes arise from alternative views of the goals and purposes of conservation. Most basic research papers on biodiversity claim to be germane to conservation practice. Many have experimental designs that are meant to mimic reductions in local biodiversity, an issue that is clearly central to conservation. However, conservation ecologists disagree on the extent to which conservation practices should be guided by scientific results, ethical considerations, or a preference to maintain organisms and ecosystems in a "natural" condition. We begin our chapter by briefly discussing practical and philosophical approaches to conservation. We then focus primarily on scientific issues, emphasizing recent research that has tested the relationship between the functioning of ecosystems and biodiversity. We emphasize how advances presented in this book and elsewhere could inform conservation practitioners. The reader should recognize that experiments on biodiversity and ecosystem functioning are few in number and for the most part relatively recent. Any single study in this area is unlikely to make an immediate contribution to practical conservation biology in the same way that a more narrow applied study might (e.g., a population viability analysis for a particular endangered species). However, in the long term this field of research should increase understanding of how communities and ecosystems work, which will help conservationists to better manage habitats and biota. We express this optimism because basic research over the last 10 to 30 years on island biogeography, metapopulation theory, the role of disturbance in ecosystems, and keystone species is now benefiting conservation efforts. It is our expectation that a deeper understanding of the role of biodiversity in ecosystem functioning will similarly advance future conservation efforts.

CONSERVATION PHILOSOPHIES
AND ECOLOGICAL SCIENCE

Preserving ecosystem functioning may provide one good reason to protect biodiversity, but it is not the only reason. Many conservationists are motivated by a deep appreciation for nature and living organisms, which they want to preserve for humanity as a natural legacy. Biodiversity loss has become a rallying point for conservation activists because it provides a scorecard for how greatly humans are making an impact on nature. However, most conservationists are more concerned with preserving natural systems than an abstract headcount of species. Similarly, some conservationists do not view preservation of ecosystem processes as an end in itself, but instead are interested in preserving processes because they are integral to preserving natural communities. Conservation efforts have often been guided by practical considerations, such as preserving imminently endangered species or land targeted for development. More recently, many conservation organizations are emphasizing ecosystem management, the role of species within ecosystems, and the sustainability of ecosystem functions. This change in approach stems from the growing concern that some of the species or habitats that have been protected will continue to decline if the area preserved is too small or otherwise inadequate to preserve the ecosystem properties upon which the biota depend (e.g., Walker 1995; Carroll et al. 1996).

While many conservationists deal with practical considerations, most also justify the preservation of species on ethical grounds. Two schools of thought, the "biocentric" conservationists and the animal rights movement, seek to protect species based on the moral responsibilities of humans to the biota, while other conservation philosophies are based on moral duties to humans. These views are not mutually exclusive. The biocentric view holds that humans are members of an interdependent biological community, and that humans should refrain from harming fellow members (Rolston 1988; Callicott 1998a). Although this ethical issue is distinct from practical issues like preserving ecosystem functioning, proponents of this ethical view believe that humans can help prevent harm to the biota by preserving ecosystem functioning. For example, Aldo Leopold (1949), who originated the biocentric view, buttressed his ethical philosophy with the then-contemporary ecological theory that organisms in biotic communities are strongly interdependent. He pointed out that human exploitation can impair ecosystem functions that are necessary for the survival of humans and other organisms (e.g., water cycling, soil fertility). He argued for preserving both biodiversity and ecosystem functioning because they are interreliant. The animal rights movement also holds that humans have a moral duty to avoid harming or exploiting animals, but on the basis of the animals' sentience, capacity to suffer, and desire for freedom; these characteristics also form the basis by which humans claim free agency and freedom from harm (review: Callicott 1998b). Without critiquing the validity of this philosophy, we point out that this is not a very comprehensive basis for conservation because species that do not meet the criteria may not be protected, and protection of one species may even endanger others (Harrop 1999).

Another ethical view is espoused by conservationists who assert that our primary moral duties are toward other humans: we should be conscientious stewards of resources for each other and for future generations. These conservationists seek to understand the various benefits that society receives from nature, including benefits derived from the use of other species and the services provided by ecosystems (Daily 1997). Such benefits are economic values as well as aesthetic and spiritual values. In this view, the only conservation policies likely to be sustainable by human society are those based on the costs and benefits society receives from nature and its conservation. The benefits received by a few individuals through harming nature must be weighed against the total costs experienced by current and future generations. Hampicke (1994) points out that if humans value nature, whether for intrinsic worth, aesthetic value, or future economic sustainability and gain, it is immoral to harm nature, especially when such harm can be avoided at reasonable cost. Myers et al. (2000) observes that humans could preserve a large fraction of earth's biodiversity for the cost of two *Pathfinder* missions to Mars.

Only a subset of conservationists follow a single, formally articulated philosophy. People support conservation for multiple reasons that are related to the philosophies above: they identify with sentient fauna like dolphins and wolves; they enjoy the beauty and services of nature and wish to preserve it for themselves and posterity; they believe that destroying species is inherently immoral; they wish to prevent negative human impacts to the environment that harm human health, other organisms, and economic sustainability.

Scientific study cannot inform every approach to conservation, but it can provide useful guidance and information for many practical conservation issues (e.g., Simberloff 1988; Hobbs and Huenneke 1992; Caughley 1994; Armesto et al. 1998; Holmes and Richardson 1999). As examples, scientists can quantify human impacts on species, elucidate the ecological and evolutionary relationships and requirements of the biota (providing the "how to" of conservation), and can characterize critical interdependencies such as the effects of human actions on the biosphere (e.g., global warming). This book focuses on the contributions of the biota to the functioning of ecosystems, and develops theory that could help conservationists understand how changes in species richness within communities will potentially feed back through ecosystem functions to cause further changes.

STUDIES OF BIODIVERSITY–ECOSYSTEM FUNCTIONING RELATIONSHIPS: ORIGINS AND RECENT CRITIQUES

Ecologists have long studied the relationship between species diversity, community stability, and ecosystem functioning (review, Tilman et al., chapter 3). Research on the role of species in ecosystems could be conducted as a purely academic exercise, but in practice, discussing the conservation implications of this work is a venerable tradition. For example, in 1958 Charles Elton observed that pest outbreaks and invasive species occur most often in communities that have been simplified by cultivation or insecticides. Elton also hypothesized a relationship between species diversity and population stability, and made it extremely clear that the diver-sity–stability relationship was important to both economics and conservation. His prime example of the value of diversity was the benefit to be gained by conserving the British hedgerow. By virtue of the variety of species they support, hedgerows supply important services, including pest control (by supporting predatory birds and arthropods), windbreaks that reduce evaporative water loss (equivalent to increasing rainfall by up to 33%), shade for livestock and humans, plus goods such as wood and berries, and beauty, which humans value. Elton called for ecologists to investigate the role of community diversity in providing such goods and services, and urged humans to carefully consider the costs they incur when decreasing the diversity of any ecological community.

Although many ecologists heeded Elton's call, early efforts to establish a relationship between stability and diversity were occasionally criticized because they were based largely on correlational data, or on mathematical models that were often considered unrealistic (Goodman 1975). Experimental studies were few, possibly because few ecologists work at the interface of community and ecosystem ecology. Fortunately the subject is now enjoying increased attention. In the early 1990s, a symposium on biodiversity and ecosystem functioning (Schulze and Mooney 1993) was followed by two influential experimental studies that suggested a relationship between species richness of field or grassland communities and the production of biomass (Naeem et al. 1994, 1995; Tilman and Downing 1994). Recent studies of microbial organisms have found similar relationships (McGrady-Steed, Harris, and Morin 1997; Naeem and Li 1997). Schmid, Joshi, and Schläpfer (chapter 6) reviewed these and other recent studies, finding that many showed a quantifiable, positive relationship between biodiversity and some ecosystem functions (especially production of biomass) and the temporal stability of these functions ("stability" *sensu* Tilman and Lehman, chapter 2).

Because human endeavors rely on ecosystem functioning (e.g., Ehrlich and Ehrlich 1992; Daily 1997), these findings strengthened the argument that we need to conserve biodiversity. However, some conservationists are understandably wary of using theory that has not been tested extensively to justify conservation programs, in case the overall relationship between diversity and functioning turns out to be weak or true for only a limited number of ecosystems and processes (Schwartz et al. 2000). Certain empirical findings could even be interpreted in a way that harms the cause of conservation. For example, the limited research to date shows that when a given ecosystem function increases with biodiversity, this effect may only occur over a small range of species richness and often appears to reach an asymptotic limit once local sites contain 10 to 15 species (Tilman et al. chapter 3 herein; Schwartz et al. 2000). This raises the possibility that some species could be considered expendable with regard to ecosystem functioning (Lawton and Brown 1993; Lawton 1994; Gitay, Wilson, and Lee 1996), and begs the question of the number of species needed to maintain full ecosystem functioning in natural or managed ecosystems. This point is addressed below and in chapter 9 by Kinzig and Pacala.

Another source of misunderstanding is that the effect of diversity on ecosystem processes and their stability has been documented in experimental communities that were assembled at random—does this mean that as long as ecosystems contain a sufficiently large grab bag of species, the identity of the species does not matter? Soule (1996) worried that if weedy species can accomplish the same ecosystem functions as natives, advocating the preservation of functioning is not a good justification for preserving intact native communities (see also Tracy and Brussard 1994; Simberloff 1998; Schwartz et al. 2000). A related challenge arises from the observation that some changes in functioning that were originally attributed to an increase in species richness might only reflect the probability of including an important dominant species in the mix, which is the sole determinant of the level of ecosystem functioning (Aarssen 1997; Huston 1997; Schwartz et al. 2000). This is known as the "sampling effect": when investigators randomly assemble species from a given species pool, the treatment with the most species is more likely to include the dominant, or critical, species for a particular process (Huston 1997; Tilman 1997a). If this were true the proposed biodiversity–functioning relationship could be spurious, and in the absurd extreme, it would only be necessary to preserve one dominant species per functional group!

This is a formidable list of shortcomings but we will discuss each of them in turn; existing studies have addressed some concerns, whereas additional research and improved experimental designs will be necessary in other cases. This field of ecology is still young, but we are cautiously optimistic that research on biodiversity and ecosystem functioning will be of great value to conservation biologists, especially those who focus on the costs and benefits to humans of alternative conservation practices.

FOUR UNRESOLVED ISSUES

(1) The relationship between ecosystem functioning and biodiversity is not well-established enough to use as an argument for preserving biodiversity.

Obviously, more research is needed on different ecosystems and on a broader variety of ecosystem functions (see Schmid, Joshi, and Schläpfer, chapter 6, for suggestions). Only a handful of experiments have been published, most of which involve either grasslands or aquatic communities in microcosms. These have found fairly consistent evidence for the diversity, ecosystem-function, stability relationship for primary productivity, as well as for some other ecosystem processes (Schläpfer and Schmid 1999; Schwartz et al. 2000;

301

Schmid, Joshi, and Schläpfer, chapter 6). Although evidence from additional types of ecosystems is needed, the discovery of comparable results in systems as disparate as aquatic microbial communities and grasslands raises the intriguing possibility that relationships between species diversity and ecosystem functioning reflect a general ecological pattern.

Studies of how natural variation in biodiversity influences ecosystem functioning have usually been consistent with theoretical predictions of a positive correlation between biodiversity, functioning, and functional stability (Schwartz et al. 2000). As examples, McNaughton (1978) and Frank and McNaughton (1991) showed that naturally more diverse grasslands can maintain primary productivity better than less diverse grasslands in the face of disturbances. Similarly, Tilman, Wedin, and Knops (1996) showed that more species-rich grassland plots showed higher productivity and better utilization of soil nitrogen than plots with fewer species. However, there are also several examples of processes and measures of stability that do not increase with diversity in natural systems (Grime 1997; Sankaran and McNaughton 1999; Schwartz et al. 2000). A complication in studies of natural gradients in diversity is that the variation in diversity may arise from disturbances or other processes that also affect ecosystem functions. Experimental studies that directly manipulate species richness come closer to testing the hypothesis that reducing species richness affects ecosystem functioning. However, these studies often fail to address the interaction of biodiversity with other key processes in nature such as disturbance. There remains a need to integrate results based on simplified experimental systems with those derived from natural systems (Grime 1997).

It is unlikely that every ecosystem process will increase with diversity, and some may decrease (Kinzig and Pacala, chapter 9; Schmid, Joshi, and Schläpfer, chapter 6). For example, Wardle, Bonner, and Nicholson (1997) found no relationship between the diversity of plant litter and its rate of decomposition. Moreover, for many ecosystem processes it is difficult to assess whether an increase or a decrease in some rate or state variable is desirable because what constitutes a "good" level of an ecosystem process is undefined. For example, it is not clear that an increase in the stability of an ecosystem process is "good" if the species in a community have become adapted to conditions that fluctuate. Judging whether changes are good or bad often involves ethical precepts that are outside the domain of science; however, quantifying the nature and extent of changes in ecosystem processes will often be essential for achieving specific management goals. An important role of science is to characterize the relationships between various ecosystem functions and diversity, and to determine the likely effects of changes in functions. One rule of thumb that may emerge for conservationists is that preserving "normal" ranges and variabilities of ecosystem processes may help preserve the native species that have become adapted to these conditions; this could help prevent further species loss and ecosystem disruption (e.g., Walker 1995).

Not only have few ecosystem processes been studied, but the range of experimental manipulations has been restricted. Most studies have considered the effects of decreased biodiversity, but humans also increase biodiversity above historical levels. For example, humans may harvest a dominant species, allowing competitive release of others. Local biodiversity has also increased in many communities because of the introduction of exotic species. Anthropogenic increases in biodiversity could also affect ecosystem functions, and some of these impacts may be considered negative. It is especially important to determine the effects of anthropogenic changes in diversity, including responses to exotic invaders, if we are to fully understand human impacts on natural ecosystems.

Schwartz et al. (2000) point out that most experimental studies to date have been biased in favor of detecting an effect of species richness by creating communities with unnatural "evenness," where equal numbers of individuals of each species initiated communities. Schmid, Joshi, and Schläpfer raise the same point in chapter 6. Natural communities are usually dominated by a small set of species, so it is questionable whether ecosystem functioning would be as strongly or immediately related to species richness in nature as in the experiments (Grime 1998). It is important to research communities with more natural structure, but at least two studies suggest that this problem may not be as serious as feared. Microbial microcosm experiments have revealed strong relationships between species richness, ecosystem functioning, and functional stability after many generations, during which dominance relationships became fully developed (e.g., McGrady-Steed, Harris, and Morin 1997; Naeem and Li 1997). In addition, the strength of the species richness–ecosystem functioning relationship increased over time in the Minnesota biodiversity study and the BIODEPTH project, where dominance relationships became more pronounced over time (Tilman et al., chapter 3; J. Joshi and A. Hector, personal communications).

(2) Ecosystem functioning forms an asymptote with species richness at a relatively low number of species.

The authors of this book have addressed this concern in several ways. Tilman (1999a) argues that the value of the asymptote is misleading about the actual species richness needed to maintain ecosystem functioning because an asymptote observed in a small plot neglects the diversity required regionally to generate a particular level of local species richness. The BIODEPTH project showed that species richness and ecosystem functions were correlated in small-scale plots of just two square meters, but the relationship weakened as more species were added, possibly approaching an asymptote near 16 species (Hector et al. 1999, chapter 4 this volume). However, such small plots would not be able to support nearly this many species if they were isolated from the surrounding community. Ecologists have long known that the number of species increases with habitat area, and that isolated "islands" of habitat are species-poor (e.g., MacArthur and Wilson 1967). Tilman used his Minnesota prairie system to calculate the number of species needed in a square kilometer to obtain a local species richness of 16 plants per square meter—it is about an order of magnitude greater at 127 to 270 species. Consequently, even if a local asymptote occurs in the functioning–richness relationship at around 16 species in 2-meter plots, it can be shown that many more species are needed to maintain normal ecosystem processes across a square kilometer. This issue is discussed further in Kinzig, Pacala, and Tilman (chapter 14).

The discussion above raises the issue of spatial scale. Problems in extrapolating across scales are well known in ecology (e.g., Levin 1992; Dutilleul 1993; MacNally and Quinn 1998; Chesson 1998). One of the information gaps in the study of biodiversity and ecosystem functioning is whether the spatial scale of the experiments is appropriate for assessing ecosystem functioning and functional stability. Some components of stability are known to be sensitive to the scale of measurement, such as density-dependent population regulation (Ray and Hastings 1996; Chesson 1998). It is therefore important to assess variation in functional stability across spatial scales. Small-plot and microcosm studies may not accurately predict ecosystem functioning at the landscape level (Schindler 1998). Small-scale studies clearly cannot tell conservationists exactly how much area requires protection to preserve ecosystem functioning. Although few large-scale data exist, species richness is likely to become increasingly important to ecosystem functioning as spatial scale increases because environmental conditions change across space (e.g., soil, microclimate, topography, hydrology), and a variety of species with different habitat tolerances would be required to maintain ecosystem functioning in this physical mosaic (Beare et al. 1995; Tilman, Lehman, and Bristow 1998; Tilman et al., chapter 3).

Chesson, Pacala, and Neuhauser (chapter 10) and Yachi and Loreau (1999) use models to demonstrate an analogous effect of temporal scale—as environmental conditions change over time, a wider variety of species will become important in maintaining ecosystem functioning. This occurs because species respond differently to abiotic conditions, and each species is unlikely to contribute the same amount of functioning under all conditions. A related argument calls for preserving a wide variety of species, genetic variation within species, and landscapes hospitable to geographic changes in species distributions, so that species will retain the evolutionary flexibility and ecological mobility necessary to sustain ecosystem functions under changing environmental conditions (e.g., Crozier 1997; Callicott, Crowder, and Mumford 1998; Chapin et al. 1998). In light of expected changes in global climate, there is lively debate about whether it is possible to predict which species will become important in the future. Many scientists feel that such predictions cannot be made with much confidence, so the wisest course is to preserve as many species as possible: an "insurance" strategy (e.g., Myers 1996; Gitay, Wilson, and Lee 1996; Edwards and Abivardi 1998; Goldstein 1999; but see Holdgate 1996). Some empirical studies have shown that species that are apparently functionally redundant under some environmental conditions may have quite different contributions to processes under altered conditions (Beare et al. 1995; Jaksic, Feinsinger, and Jimenez 1996; Walker, Kinzig, and Langridge 1999; Sullivan and Zedler 1999). In such cases the concept of redundancy should be replaced with the engineering idea of fail-safes—additional features that allow functioning to be maintained under a variety of situations (Naeem 1998).

(3) Because randomly assembled communities show that biodiversity influences ecosystem functioning, species identity may be unimportant.

We would like to make two comments on this issue. First, for many existing studies, the premise underlying this criticism is erroneous. Species may have been grown together in random combinations for particular experiments, but they were usually drawn from pools of sympatric species that one could reasonably expect to interact based on their common history (e.g., grasslands in the U.S.A. or Europe, Tilman and Downing 1994, Hector et al. 1999; early successional fields in England, Naeem et al. 1994). Organisms in the microcosm studies were perhaps chosen more haphazardly (McGrady-Steed, Harris, and Morin 1997; Naeem and Li 1997), but aquatic microorganisms are very widely distributed and can be expected to interact with virtually any other species (Finlay, Maberly, and Cooper 1997; Finlay and Esteban 1998). Second, randomized communities could be used as an informative treatment because the effect of human perturbations is often the disruption of communities through extinctions and introductions of exotics. There is growing evidence that communities are structured by ecological sorting of species into compatible combinations, via well-documented mechanisms like competition, predation, and mutualisms, or through evolutionary changes. Tilman and Lehman's models (chapter 2) demonstrate that competitive interactions can help to stabilize ecosystem functions (see also Tilman, Lehman, and Bristow 1998). Thompson (1999) reviewed evidence that species often show locally coevolved interactions. Extinctions and introductions could disrupt such interactions, potentially entailing a cost in ecosystem functioning above that expected in noninteractive assemblages, because the contribution to ecosystem functioning of each species is more likely to be dependent on others.

(4) Dominant species may be solely responsible for correlations between biodiversity and ecosystem functioning.

Huston (1997) and others (e.g., Hooper and Vitousek 1997) pointed out that dominant species or particular functional groups can contribute disproportionately to ecosystem functioning, leaving little room for an additional effect of species diversity. It is clear that all species are not equal in terms of their contribution to a particular ecosystem process at any given time (see also Walker 1995; Holdgate 1996; Myers 1996; Hooper and Vitousek 1997; Grime 1997, 1998). However, different species may emerge as dominants as environmental conditions change, therefore a diverse suite of potential dominants may be required in a local biota in order to tolerate a wide range of environmental situations. In addition, as Tilman et al. show (chapter 3), some species mixes can increase ecosystem functioning markedly above that expected, based on monocultures of dominant species (e.g., overyielding of biomass). More research is necessary to discover which types of species most influence various ecosystem processes, and to what degree species interactions contribute to ecosystem functioning.

RELATING BIODIVERSITY THEORY AND EXPERIMENTS TO LOSSES IN BIODIVERSITY CAUSED BY HUMANS

It should be obvious to any observer of the modern world that humans do not imperil species at random (Purvis et al. 2000). As Chapin et al. (1998) have said, "there are clear winners and losers among species as a result of human activity." As biodiversity theory and research advances, it will be made more relevant to conservation if it examines the particular consequences of the types of biodiversity losses caused by humans. Below, we provide a list of the types of species that humans tend to harm the most, and inspection of this list indicates that these species may often be among

308

those that contribute most significantly to ecosystem functioning. We list ways in which humans remove species from ecosystems in approximate order of expected size of effect on ecosystem functioning. This list may provide fruitful areas for future research.

- *Removal of dominants* (e.g., logging of forest canopy trees). Dominants are species that have large effects in ecological communities by virtue of their biomass or abundance (Power et al. 1996). This book and other work suggest that dominants often accomplish the bulk of ecosystem function, thus their removal is likely to have extreme effects on systems (review: Schmid, Joshi, and Schläpfer, chapter 6).
- *Removal of top predators* (e.g., wolves, tuna, predaceous insects). Top predators may act as "keystones" in communities, whose effects far exceed that expected by the amount of biomass or diversity removed (Mills, Soule, and Doak 1993; Power et al. 1996). Their removal has been associated with a variety of strong effects, including trophic cascades that dramatically alter ecosystem functioning (e.g., Carpenter et al. 1987), loss of diversity in other trophic levels (Paine 1966), and crop pest outbreaks leading to production declines in agriculture and forestry (DeBach 1974).
- *Reduction of diversity through pollution.* This often impacts entire functional groups (e.g., siltation can reduce algae and stream insects by blocking light and clogging gills), or it may remove species more haphazardly.
- *Nutrient enrichment, leading to loss of species adapted to low-nutrient conditions.* This has been demonstrated to reduce the ability of communities to maintain ecosystem functioning in changing environmental conditions (e.g., Tilman and Downing 1994). It may also produce changes in dominance and loss of functional groups (Vitousek et al. 1997).

In addition to the list above, humans may overharvest or purposefully destroy a variety of species that are regarded as delicious, attractive, medicinal, or noxious. It is difficult to classify the likely effects of removing these species because they possess a mixture of functional properties. In many cases, species are overharvested because they have unusual characteristics. This may sometimes be correlated with a greater likelihood of having unique or rare contributions to ecosystem functioning. Rare species are also lost disproportionately because of human activities, since small populations are poorly buffered against habitat change, loss, or disturbance. Rare species are perhaps less likely than other species to contribute much to current ecosystem functions, but could become important under changed environmental conditions. Rare species may maintain ecosystem functioning in rare habitats, and some rare species may be "keystone species" that have effects disproportionate to their biomass (Power et al. 1996).

Where Should Biodiversity Research Move in the Future If It Is to Best Address Conservation Problems?

To a certain extent, biodiversity research will be most useful to conservation if it substantially advances ecological theory and our basic understanding of ecosystem functioning. But such a platitude offers little in terms of setting priorities or raising challenges. Below, we list research topics that go beyond simply demonstrating that biodiversity can affect ecosystem functioning, and address more specific questions that could aid in preservation and management of ecosystems.

1. We need research that builds tools for predicting what sorts of ecosystem failures or collapses are likely with the loss of particular species or functional groups of species. Even more challenging is learning whether there might be signals in the dynamics or processes of communities that foreshadow the future harm associated with species loss, so

that action can be taken quickly. For example, hidden in Dave Tilman's prairie plant data, are there indicators that would have allowed him to predict the consequences of the severe drought of the 1980s before it happened (as opposed to interpreting it after the fact)?

2. Population viability theory has developed many tools for considering how large a population must be to remain viable, how many subpopulations are needed, and to what extent the surrounding habitat influences the viability of a protected population. We are starving for a parallel theory that applies to communities. Most on-the-ground conservation is currently aimed at preserving communities and landscapes (Groves et al. 2000). In deriving plans for land use and land protection, the big unanswered question is how we can tell whether the systems we are aiming to protect are sustainable.

3. The role of altered disturbance regimes (especially fire) and exotics cannot be left out of biodiversity–ecosystem function theory or experiments. In many communities, anywhere from 10–40% of the species are alien species (Vitousek et al. 1996); for example, the introduction of alien plants has increased California plant diversity from 4,839 to 5,862 species, which is greater than a 20% increase in diversity (data from Jepson 1993). These figures regarding invasive species dramatize the limitations of any biodiversity theory that neglects the role of exotic species.

4. Instead of asking what the evidence is that biodiversity matters, mathematical models and experiments need to begin asking: What circumstances (community structures, disturbance regimes, spatial scales) make losses of diversity most likely to have major impacts?

5. Mathematical models relating biodiversity to ecosystem function vary greatly in their value. Models that are analogies are of less value than more classical models of species interactions that can reveal mechanism. Of course, ecology has moved beyond its reliance on the simplest Lotka-

Volterra systems of equations, and the need for models that explicitly represent nutrient use, response to variable environments, and trophic relationships is great.

Do Conservationists Need the Results of Biodiversity Experiments to Justify Their Work?

The bulk of this chapter has dealt with evidence linking biodiversity with ecosystem functioning. An appreciation for the functional value of biodiversity (in addition to other values) can only aid the cause of conservation. If conservationists do not establish some tangible or economic value for species, other parties may simply assume that the value is zero (Myers 1996; Edwards and Abivardi 1998). It is, of course, extraordinarily difficult to come up with hard numbers (Costanza et al. 1997). However, one can prove the existence of enormous tangible value by recognizing that the world's economy is almost completely dependent on ecosystem functions and their stability, and nearly all species have a potential role in supporting these processes (e.g., Hampicke 1994).

While it is important to establish that ecosystem functions (and therefore species) have vast economic value, this can never be the only reason to conserve biodiversity. Elton (1958) certainly argued that people should consider the tangible benefits of diversity, but his first and most eloquent appeal to the reader for preserving hedgerows was on the basis of aesthetic value: "There is extreme pleasure for the traveler—the flowering hawthorn hedge and its associated shrubs like dogwood and elder, the roadside flowers and the insects that frequent them—like the brimstone butterfly. . . ." All species have aesthetic values that make them worth preserving—they have beautiful forms, colorations, and morphological designs, plus intricate behaviors and natural history that we find fascinating. Species also possess a plethora of biologically active compounds that may provide cures for diseases of humans, livestock, and crops. We have an ethical responsibility to preserve aesthetic, medicinal, and economic values for future generations (Hampicke 1994), and many people feel that we also have a responsibility to avoid harming other living creatures whenever possible.

To conclude, preserving reliable ecosystem functions may prove to be an excellent reason to conserve biodiversity locally, regionally, and globally. However, more studies are necessary to characterize the relationships between diversity, particular functional groups, and various ecosystem functions. While biodiversity experiments are not intended to provide a justification for protecting particular species or ecosystems, they are important to conservation because they provide information needed to protect and maintain viable communities. Studies of biodiversity and the interplay of community structure and ecosystem function are the cornerstone of applied conservation, even when the motivation for preserving nature is spiritual, ethical, aesthetic, or economic. This is because biodiversity studies will be crucial to understanding how communities and ecosystems function, and conservation will not succeed without such knowledge.

ACKNOWLEDGMENTS

We thank several people for their thoughtful comments on this chapter: A. Kinzig, D. Tilman, A. Hector, M. Holyoak, M. Greaves, D. Piechnik, K. Leyse, and S. Harrison. J.J.A. was supported by an endowed Presidential Chair in Science.

CHAPTER FOURTEEN

Looking Back
and Peering Forward

Ann P. Kinzig, Stephen Pacala, and David Tilman

In this volume, we have summarized the empirical evidence for a connection between species diversity and ecosystem functioning. We have been able to extend those empirical demonstrations with theoretical analyses that aid in understanding when and under what circumstances alternative forms of the diversity–functioning relationship might emerge. In addition, the theory allows us to extend experimental results to larger spatial and longer temporal scales. We hope that this volume will help resolve some of the recent controversies that have characterized the scientific debate on diversity–functioning relationships, although this book was not originally conceived in that vein. Yet we harbor no illusions that this volume will resolve all controversies, and it may even introduce new ones. That is as it should be, for it is often controversy that forces reevaluation of our dominant paradigms and theories, and skepticism that forces us to deepen or change our understanding of ecological systems.

In nearly all cases examined in this volume—both theoretical and empirical—biodiversity per se appears to influence ecosystem functioning. In the majority of cases, that influence appears to be positive—that is, increasing biodiversity increases the processes of primary production and carbon storage while decreasing nitrogen leaching. In a few cases, the opposite relationships emerge—increasing bio-

314

diversity can actually decrease primary production and other ecological processes. This is particularly true in systems where diversity is maintained by successional niches, or, in some cases, by trophic or antagonistic interactions. More specifically, we have found that:

• It is crucially important to separate community composition and species number in interpreting experimental results. The influence of the two forms of diversity on ecosystem processes can vary across sites, and across space and time within sites.

• It is also important to distinguish the sampling effect from complementarity effects when interpreting the results of experiments. In heterogeneous environments, however, such as those discussed in chapters 9 and 10, the clear distinction between sampling and complementarity that exists in homogeneous environments may be lost, because local superior competitors may vary from site to site, or may fail to dominate at the landscape level.

• In spite of these difficulties, long-term experimental results are generally consistent with an increase in functioning with diversity driven by niche-complementarity mechanisms—that is, an "overyielding," where the best polycultures outperform the best of the more depauperate communities (chapters 3 through 6). On the theoretical side, the spatial and temporal heterogeneity niche models show results consistent with these outcomes (chapter 10), as does, under certain circumstances, the competition-colonization model (chapter 9). Other classes of resource-based niche-complementarity models would be expected to exhibit these relationships as well.

• There are, however, some crucial experimental and theoretical exceptions to this general result, where either average functioning declines with diversity and/or the best polycultures have lower performance than the best monocultures (chapter 6). Theoretical analyses suggest that these outcomes could emerge (though need not emerge) when successional niches (chapter 9), trophic interactions (chapter 11), or antagonistic interactions (not covered in this volume) are the mechanisms maintaining species coexistence.

• In spite of the evidence for niche complementarity in the early experimental record, nearly all *early* experimental results were also consistent with a "sampling" hypothesis—on average, an increase in functioning with diversity, but competitive dominance by the best types in polyculture that leads to a constant upper bound on performance across diversity levels. We now know, based on analyses presented in this volume and elsewhere, that we should have been able to anticipate these results. The dynamics of communities far from equilibrium—like those communities initially established in experimental plots—are characterized by two or more distinct time dynamics: a relatively rapid early growth of existing individuals—particularly "weedy" species—and the slower dynamics associated with shifts in abundance and composition. The early dynamics influence diversity–functioning relationships through sampling, and the later dynamics influence diversity–functioning relationships through niche complementarity mechanisms (chapter 7).

• Our knowledge of the influence of microbial diversity on ecosystem functioning is still restricted. This situation is in part due to the limitations of methods for adequately measuring microbial diversity, and relating those measures to ecosystem processes. Nonetheless, our analyses (chapter 12) suggest that microbial diversity per se is most likely to exert an influence on ecosystem functioning when (1) ecosystems are undergoing a transition from one equilibrium to another; (2) the ecosystem process of interest is carried out by only a few types of microorganisms; (3) these transitions are caused by changes in environmental factors, rather than changes in resource levels; and (4) perturbations are visited upon systems that had heretofore experienced a relatively narrow range of environmental conditions.

• Most experiments are unable to span the spatial and temporal scales, or the trophic structures, that are relevant for a thorough understanding of diversity–functioning relationships. The proper role of theory in this field is to extend the experimental results to these other spatiotemporal scales and community structures, and to elucidate the impacts of different coexistence mechanisms on diversity–functioning relationships. Experimental manipulations or field observations will, in turn, be required to validate the theory. We elaborate on these points, and offer some suggestions about future directions for research in this field, in the remainder of this chapter.

Ecologists have made enormous strides in elucidating the diversity–functioning relationship since the publication in 1993 of the seminal volume edited by Schulze and Mooney. We have summarized some of those advances in this volume. Nonetheless, there are significant conceptual and empirical advances that still need to be made if we are to better understand how ecosystems are structured and function, and to aid the public and policymakers as they struggle to balance the integrity of biotic systems and ecosystem services against other societal benefits that can derive from use and conversion of natural resources.

What are the biggest remaining gaps in our knowledge? First and foremost, natural communities are neither randomly assembled nor randomly disassembled. Both theory and experiment have proceeded by establishing all possible n-species communities (or a statistically reasonable subset of them) drawn from an N-species pool. The N-species pool usually represents the extant regional diversity in the system, though it may in some cases exclude very rare species. But just because an n-species community can exist does not mean it will, particularly when it exists under the conditions of an experimental manipulation that relies on weeding and other interventions to maintain the community.

By conducting the draw in this way, ecologists are creating a situation in which each n-species combination is equally likely, and awarding each equal weight in determining the diversity–functioning relationship. This is an efficient way to separate the pure effects of diversity from effects attributable to correlated changes in diversity and composition. This experimental design can give sufficient statistical power to distinguish the impacts of species identity from species diversity, i.e., is it diversity per se, or the identity of a few key species that determines the level of functioning of an ecosystem? Given the complexity of interactions in the system, experimental designs that draw on numerous species combinations are warranted. As for the theoretical analyses, drawing species randomly from an N-species pool allows determination of the pure effects of diversity, but not the effects of community assembly, which may well cause biases in composition as communities assemble. The models of coexistence used in this monograph are not necessarily models of assembly. Given an extant regional pool, the coexistence models can suggest which species can coexist locally, but not necessarily which species are likely to be found together. Nor do current coexistence models adequately address the question of what determines the make-up of the regional species pool to begin with.

The filters that Nature applies to determine which species will coexist in a square meter plot are—given what we know about competitive displacements and limiting similarity— likely more sophisticated than a random draw. Two very similar species may appear together in the regional pool, and yet only rarely or never appear together locally. Local co-occurrence at a frequency greater than that implied by a random draw could also occur. Are the communities that Nature assembles, then, likely to have greater functioning than communities of similar diversity constructed by random draw? Lesser functioning? Will their range of functioning span that found in the random draw? Any answer offers interesting insights into the assembly rules that determine local communities, and how those assembly rules—which presumably rely on small-scale interactions among individuals, as well as dispersal and recruitment limitations—influence the emergent properties of ecosystem functioning, for which there is no direct selection.

We also have not yet been able to elucidate what level of regional diversity is required to maintain the levels of local diversity needed to sustain functioning. Experiments are conducted on patches of 1 to perhaps 100 square meters. These are sensible patch sizes for measuring functioning in the grassland systems that have largely been the subject of experiment. It is over scales on the order of a square meter, after all, that most grassland species interact, and thus over these scales that species interactions could be expected to influence functioning. (Needless to say, the relevant patch size will vary from system to system, and is defined by the characteristic scale over which individual interactions occur. But the relevant scale for measuring functioning will always be substantially smaller than the extent of the ecosystem or biome that will be the focus of management and conservation efforts.)

We know from species–area relationships that regional diversity is always greater than local diversity (as it must be)—how much greater depends on the sizes of the areas under consideration, and biomes and taxonomic groups in question. But these areal correlations do not necessarily provide information on the regional diversity required to maintain a given level of local diversity and functioning. We may find that removing species from the regional pool leads to a proportional decline of species from the local pool—as we would expect if the species–area relationship were indicative of a causal rather than correlational relationship between regional and local diversity. On the other hand, we may find that we can remove certain species from a regional pool and still maintain local diversity—functionally similar species may merely replace, at the local level, those lost from the regional pool.

But these are precisely the questions that are of interest to the public and policymakers. What level of regional diversity is required to maintain local and regional functioning? If we lose 10% of the species regionally, what is the degradation in functioning? How does it depend on which species are lost, and which species are likely to be lost? To answer those questions, we would need to know something about the level of local diversity required to maintain functioning, the level of regional diversity required to maintain local diversity, and, likely, the identity of the species lost from the regional pool. These are the crucial connections that we are, as of yet, unable to make as mechanistically as we might wish.

How might we get there? Consider the series of graphs in figure 14.1, which show the relationships between productivity (Prod), regional experimental diversity (R_E), local experimental diversity (L_E), regional field diversity (R_F), and local field diversity (L_F). (By "field" diversity, we mean the diversity levels found—either locally or regionally—in the nonexperimental, nonmanipulated system that the experiment purports to simulate.) What we have found thus far via experiment is encapsulated in figure 14.1A, the relationship between regional experimental diversity and productivity (taken here as an example of functioning). What we would like to achieve is represented in figure 14.1F—the relationship between regional field diversity and productivity. Making our way from 14.1A to 14.1F will mean taking several steps in between.

Consider first figure 14.1A–C. Experimental results of functioning are reported relative to the n-species pool (out of a total possible pool of N) used to initially establish the community. The n species are planted in each plot, and any invaders not from that pool that appear in the community are removed. Thus, the local community that does establish at any particular diversity level does so under a rain of seeds from n species—equivalent to a regional pool of n species. (We are, of course, ignoring recruitment limitations here—all n species visit the local site simultaneously. The local community is thus established from a regional pool whose members do not suffer any recruitment limitations.) Experimental results, therefore, show the relationship between R_E and production (figure 14.1A).

Not all of the species initially planted, however, will necessarily persist indefinitely in the local communities. Thus, an n-species regional pool might lead to a more depauperate local community. The relationship between local diversity and the regional pool of n may or may not be easy to obtain from the experimental record—many experiments are, after all, of shorter duration than the dynamics of competitive interactions that might eventually lead to local exclusion of an originally planted species. One might obtain an estimate, however, by assuming that any species showing a substantial decline in abundance over the course of the experiment—or that is very rare by experiment's end—is a likely candidate for eventual local extinction. This would then give a relationship between L_E and Prod, as shown in figure 14.1B. (Note that these figures are conceptual, and not drawn from real data.) Finally, it would also be easy enough, from the experimental data, to figure out the relationship between R_E (the regional n-species pool used to establish the community) and L_E (the resulting local species number) (figure 14.1C).

How do these results inform our efforts to find the equivalent field-level relationships (figure 14.1D–F)? The observed relationship between R_E and L_E might offer some insights into the relationship between R_F and L_F but, as noted above, the local experimental communities are established in the absence of recruitment limitations. Recruitment limitations would only serve to depress the number of local species that could be maintained for any given regional pool.

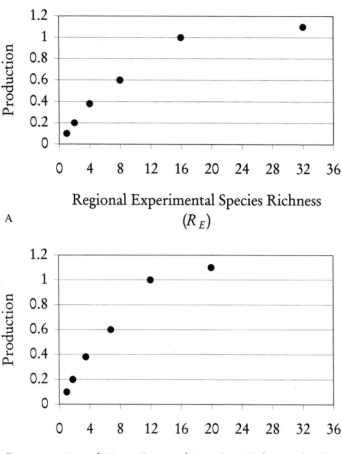

A

B

FIGURE 14.1. The steps required to translate local experimental results to an understanding of regional diversity–functioning relationships. See text for a more detailed explanation. (A) Hypothetical experimental results indicating the relationship between regional experimental diversity (the *n*-species pool used to seed the plots) and productivity (as one possible example of functioning). (B) Hypothetical relationship between local experimental diversity (number of species remaining in plot after initial seeding) and productivity.

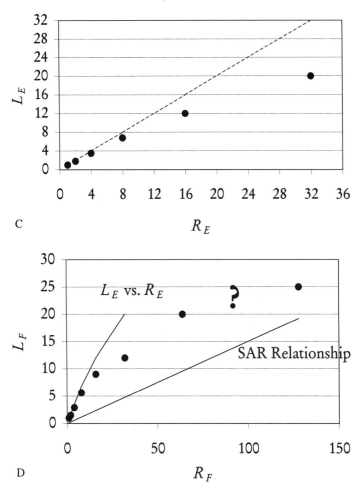

Figure 14.1 continued: (C) Hypothetical relationship between regional and local experimental diversity. The number of species remaining in a plot will be less than or equal to the number used to seed the plot (assuming invaders are eliminated). (D) Potential relationships between regional and local field diversity (defined as the extant species in natural rather than experimental communities). Upper and lower bounds are given by experimental relationships (fig. 14.1C) and species-area relationships (SAR), respectively. The actual relationship is likely to fall within these two bounds. See text for further details.

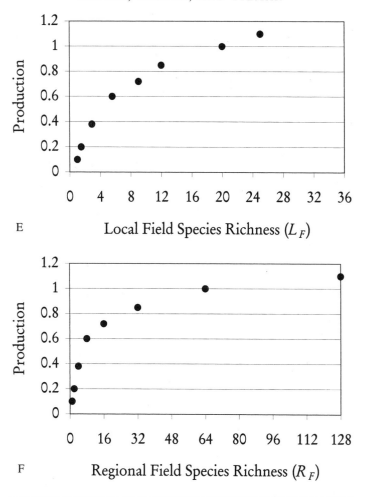

Figure 14.1 continued: (E) Hypothetical relationship between local field (as opposed to experimental) diversity and functioning. (F) Hypothetical relationship between regional field diversity and productivity.

Further, the L_E's as given in the experiments may be inflated because species are "hanging on" through the initial, transient stages of the experiment but destined to disappear from the local community under competitive pressures at a later time. Thus the L_E as a function of R_E relationship only serves as an upper bound for the equivalent field relationship between R_F and L_F (figure 14.1D).

An alternative approach is provided by the species–area relationship (Tilman 1999a). The species–area relationship states that the number of species in a locality, S, depends on the area of the sites, A, raised to the power z (MacArthur and Wilson 1967). Let the local area, A_L, be the size of the sites within which individuals interact and thereby influence ecosystem functioning, and let A_R be the size of a region within which such sites are embedded. The species–area relationship predicts that, for a local site to have a number of species S_L, it would have to be embedded in a region with diversity of S_R, where $S_R = S_L (A_R/A_L)^z$. What does this imply? For a grassland to contain 16 species in each 1 m^2 area, even a low z value (e.g., $z = 0.15$) would mean that the surrounding 100 hectare region would have to contain 127 species (Tilman 1999a). If each 1 m^2 area must contain 16 species to assure maximal productivity for a grassland ecosystem, for instance, then a 100 hectare field would have to contain 127 species and a 100 km^2 region would have to contain 253 species to maintain this level of diversity, and thus productivity, within the average 1 m^2 area. The empirically observed species–area relationship implies that a decrease in regional diversity leads, in natural communities, to a proportional decrease in local diversity and thus, by extension, to potentially decreased ecosystem functioning.

The species–area relationship has been frequently observed, and a series of underlying mechanisms have been proposed to explain the link in diversity across various scales that is embodied in it (e.g., Rosenzweig 1995; Hubbel 2001). However, in one sense, the species–area relationship could be interpreted as indicating that although 127 species are sufficient to maintain 16 species locally, a smaller regional pool might be able to maintain this same level of local diversity. Removal of some of the species from the regional pool may not necessarily decrease local diversity in the long term. If some species are functionally similar, for instance, complementary species might replace, at the local scale, those species lost from the regional pool, and thus prevent any long-term degradation in local diversity or functioning. The species–area relationship might therefore give a lower bound on L_F as a function of R_F—at most we would need the number of species present in the regional pool to maintain local diversity, but we may need fewer (figure 14.1D).

In some systems, these upper and lower bounds may be fortuitously close together, allowing identification of the true L_F–R_F relationship within a fairly narrow range. More likely, the curves will lie farther apart, necessitating further experimentation or analysis to discern how degradations in regional diversity actually alter local diversity. One might assume that systems that have significant recruitment limitations and little functional redundancy would lie closer to the lower bound determined by species–area relationships. The absence of functionally redundant species in the regional pool means that any loss will reduce local functioning somewhere. Systems that have little recruitment limitation, as well as substantial functional redundancy, might lie closer to the upper bound determined by experimental relationships between local and regional diversity. Functional redundancy and low recruitment limitations allow lost species to be replaced, with little time lag, by a functionally similar one. Clearly, additional experimentation may be the only way to narrow the range of plausible relationships further.

One possibility is to conduct a set of experiments that effectively remove subsets of species from the regional pool. Small plots could be established; species assumed to be lost from the regional pool would be removed from the local one. The seed-rain from these lost regional species would also have to be removed, either by weeding of invaders or trapping and sorting of seeds, with the seeds of those species still assumed to be in the regional pool returned to the site. Depauperation of the regional pool could occur at fairly high rates—5% to over 50% of species, for instance—in order to determine the proper L_F–R_F relationship. Note that there is no need to monitor functioning; the experiment is intended to determine diversity and not functioning relationships. Long-term monitoring and removal may be necessary. These experiments could and should be complemented with additional theoretical developments that could further elucidate the regional and local "filters" Nature uses in assembling communities.

How might we make the measurements required to create figure 14.1E? There we are interested in the relationship between local field diversity (nonrandomly assembled) and functioning. One simple approach would just be observation—establishing plots of one square meter, counting the number of species on them, and measuring functioning. But, as many have noted before, causal effects are difficult to distinguish using this approach. Is functioning higher because there is a greater number of species at a particular site, or is there a greater number of species at a particular site because conditions were initially favorable both for diversity and functioning? Frequently, there is insufficient statistical power to identify the dominant causal relationships. Thus, this approach might best be used in a fairly homogeneous environment, where there is little reason to believe that there are extant topographic, hydrologic, or edaphic conditions that might create more favorable conditions in some small-scale plots relative to others.

An alternative would be to conduct removal and addition experiments in natural communities—with the perturbations ranging from modest (addition or removal of only one or two species) to severe (addition or removal of many). Both types of experiments would need to be sufficiently long-term to overcome lags associated either with recruitment limitation following removals (some dominant species can require 30 or more years to become established during old-field succession) or with growth, maturation, and interspecific interactions following species additions. The more modest perturbations may, perhaps, be more likely to produce "realistic" local communities—that is, some collection of species that could be found together and persist together over reasonable timescales in a nonexperimental setting— but the more severe perturbations may be necessary to capture the range of influences of human activities on local and regional species pools. One would, in effect, create a "species map" for a region—tallying all of the extant n-species communities and their compositions at the local scale—and then randomly add or delete species from that map to create the experimental communities. Provisions would have to be made to correct for variations in cover among plots, and to monitor other factors (again, topographic, hydrologic, or edaphic) that may be influencing functioning.

One such provision would be to take the local communities established in the field using the above approach, and replicate them in a common garden. Edaphic and other factors could be controlled in this common garden—and differences from plot to plot minimized. The diversity–functioning relationships resulting from addition-removal experiments in extant communities could then be compared to those measured in the common-garden plots to identify the potential magnitude of the influences of variations in geophysical conditions on diversity and functioning.

As the experimental evidence for assembly rules and functioning accumulates, theorists can again step in to extend the results of short-term experiments to longer timescales, and to help identify the commonalities and differences among different biomes and sites. Creation of a "unified theory" of community assembly and disassembly across spatial scales should be the theorist's holy grail.

The world is destined to lose some species. There is no other possible outcome, given an expanding human population, and expanding consumption. Loss of species could be costly in terms of degradation of ecosystem functioning and loss of ecosystem services. But preservation of species may be costly as well—both in terms of direct expenses for land acquisition and patrolling, for instance, and in terms of the "opportunity cost" associated with setting aside parcels of land for this purpose. The decisions that societies must make about managing and preserving regional diversity need to be informed by relevant scientific knowledge. Although we have learned much about diversity and ecosystem functioning during this past decade, many questions remain unanswered. We hope this book can help inspire and focus the research needed to address these questions.

Such knowledge would make the science of diversity–functioning relationships more societally useful. At the same time, understanding how communities assemble, how they disassemble, and how diversity–functioning relationships can emerge as the result of coexistence and assembly processes that cross spatial and temporal scales is at the heart of some of the most fundamental and compelling unanswered questions in ecology. Extending our examination of diversity–functioning relationships to new approaches and conditions promises to yield substantial intellectual and societal benefits.

References

Aarssen, L. W. 1983. Ecological combining ability and competitive combining ability in plants: toward a general evolutionary theory of coexistence in systems of competition. *American Naturalist* 122:707–731.

—. 1997. High productivity in grassland ecosystems: effected by species diversity or productive species? *Oikos* 80:183–184.

Abrams, P. 1976. Environmental variability and niche overlap. *Mathematical Bioscience* 28:357–372.

—. 1984. Variability in resource consumption rates and the coexistence of competing species. *Theoretical Population Biology* 25:106–124.

Abrams, P. A. 1993. Effect of increased productivity on the abundance of trophic levels. *American Naturalist* 141:351–371.

Allison, G. W. 1999. The implications of experimental design for biodiversity manipulations. *American Naturalist* 153:26–45.

Anderson, T. H. 1994. Physiological analysis of microbial communities in soil: applications and limitations. In K. Ritz, J. Dighton and K. E. Giller, eds., *Beyond the Biomass*, 67–76. London: Wiley-Sayce.

Anderson, T.-H., and K. H. Domsch. 1993. The metabolic quotient for CO_2 (qCO_2) as a specific activity parameter to assess the effects of environmental conditions, such as pH, on the microbial biomass of forest soils. *Soil Biology and Biochemistry* 25 (3): 393–395.

Andrén, O., and J. Balandreau. 1999. Biodiversity and soil functioning—from black box to can of worms? *Applied Soil Ecology* 13:105–108.

Andrén, O., L. Brussaard, and M. Clarholme. 1999. Soil organism influence on ecosystem-level processes—bypassing the ecological hierarchy? *Applied Soil Ecology* 11:177–188.

Armesto, J. J., R. Rozzi, C. Smith-Ramirez, and M.T.K. Arroyo. 1998. Conservation targets in South American temperate forests. *Science* 279:1271–1272.

Armstrong, R. A. 1976. Fugitive species: experiments with fungi and some theoretical considerations. *Ecology* 57:953–963.

Atlas, R. M., and R. Bartha. 1993. *Microbial Ecology: Fundamentals and Applications.* New York: Benjamin Cummings.

REFERENCES

Balser, T. C. 2000. *Linking Soil Microbial Communities and Ecosystem Functioning*. Department of Soil Science, University of California, Doctoral Dissertation.

Bardgett, R. C., and A. Shine. 1999. Linkages between plant litter diversity, soil microbial biomass and ecosystem function in temperate grasslands. *Soil Biology and Biochemistry* 31:317–321

Baskin, C. C., P. Chesson, and J. M. Baskin. 1993. Annual seed dormancy cycles in two desert winter annuals. *Journal of Ecology* 81:551–556.

Bazzaz, F. A. 1979. The physiological ecology of plant succession. *Annual Review of Ecology and Systematics* 10:351–372.

Beare, M. H., D. C. Coleman, D. A. Crossley, Jr., P. F. Hendrix, and E. P. Odum. 1995. A hierarchical approach to evaluating the significance of soil biodiversity to biogeochemical cycling. *Plant and Soil* 170:5–22.

Berish, C. W., and J. J. Ewel. 1988. Root development in simple and complex tropical successional ecosystems. *Plant and Soil* 106:73–84.

Berman-Frank, I., and Z. Dubinsky. 1999. Balanced growth in aquatic plants: myth or reality? *BioScience* 49:29–37.

Bock, C. E., and J. H. Bock. 1974. Geographical ecology of the acorn woodpecker: diversity versus abundance of resources. *The American Naturalist* 108:694–698.

Bolker, B., and S. W. Pacala. 1997. Using moment equations to understand stochastically driven spatial pattern formation in ecological systems. *Theoretical Population Biology* 52:179–197.

Bolker, B. M., and S. W. Pacala. 1999. Spatial moment equations for plant competition: understanding spatial strategies and the advantages of short dispersal. *The American Naturalist* 153 (6): 575–602.

Bolker, B. M., S. W. Pacala, and W. J. Parton. 1998. Linear analysis of soil decomposition: insights from the Century model. *Ecological Applications* 8 (2): 425–439.

Borneman, J., and E. W. Triplett. 1997. Molecular microbial diversity in soils from Eastern Amazonia: evidence for unusual microorganisms and microbial population shifts associated with deforestation. *Applied and Environmental Microbiology* 63:2647–2653.

Borneman, J., P. W. Skroch, K. M. O'Sullivan, J. A. Palus, N. J. Rumjanek, J. L. Jansen, J. Nienhuis, and E. Triplett. 1996. Molecular microbial diversity of an agricultural soil in Wisconsin. *Applied and Environmental Microbiology* 62 (6): 1935–1943.

Bowers, M. A. 1987. Precipitation and the relative abundance of desert winter annuals: a 6-year study in the northern Mohave Desert. *Journal of Arid Environments* 12:141–149.

Brock, T. D., and M. T. Madigan. 1991. *Biology of Microorganisms.* Upper Saddle River, N.J.: Prentice Hall.

Brown, B. J., and J. J. Ewel. 1987. Herbivory in complex and simple tropical successional ecosystems. *Ecology* 68:108–116.

Brown, J. S. 1989. Coexistence on a seasonal resource. *The American Naturalist* 133:168–182.

Burkhardt, C., H. Insam, T. C. Hutchinson, and H. H. Reber. 1993. Impact of heavy metals on the degradative capabilities of soil bacterial communities. *Biology and Fertility of Soils* 16:154–156.

Burns, T. P. 1989. Lindeman's contradiction and the trophic structure of ecosystems. *Ecology* 70:1355–1362.

Callicott, J. B. 1998a. Do deconstructive ecology and sociobiology undermine Leopold's land ethic? In M. E. Zimmerman, J. B. Callicott, G. Sessions, K. J. Warren, and J. Clark, eds., *Environmental Philosophy: From Animal Rights to Radical Ecology,* 145–164. Upper Saddle River, N.J.: Prentice Hall.

―――. 1998b. Environmental ethics: introduction. In M. E. Zimmerman, J. B. Callicott, G. Sessions, K. J. Warren, and J. Clark, eds., *Environmental Philosophy: From Animal Rights to Radical Ecology,* 7–16. Upper Saddle River, N.J.: Prentice Hall.

Callicott, J. B., L. B. Crowder, and K. Mumford. 1998. Current normative concepts in conservation. *Conservation Biology* 13:22–35.

Carpenter, S. R., and J. F. Kitchell, eds. 1993. *The Trophic Cascade in Lakes.* Cambridge: Cambridge University Press.

Carpenter, S. R., J. F. Kitchell, J. R. Hodgson, P. A. Cochran, J. J. Elser, M. M. Elser, D. M. Lodge, D. Kretchmer, and X. He. 1987. Regulation of lake primary productivity by food web structure. *Ecology* 68:1863–1876.

Carpenter, S. R., J. F. Kitchell, K. L. Cottingham, D. E. Schindler, D. L. Christnesen, D. M. Post, and N. Voichick. 1996. Chlorophyll variability, nutrient input, and grazing: evidence from whole-lake experiments. *Ecology* 77:725–735.

Carroll, R., C. Augspurger, A. Dobson, J. Franklin, G. Orians, W. Reid, R. Tracy, D. Wilcove, and J. Wilson. 1996. Strengthening the use of science in achieving the goals of the endangered species act: an assessment by the Ecological Society of America. *Ecological Applications* 6:1–11.

Caughley, G. 1994. Directions in conservation biology. *Journal of Animal Ecology* 63:215–244.

Caughley, G., and J. H. Lawton. 1976. Plant-herbivore systems. In

R. M. May, ed., *Theoretical Ecology*, 2d ed., 132–166. Sunderland, Mass.: Sinauer Associates.

Cavigelli, M., and P. Robertson. 2000. The functional significance of denitrifier community composition in a terrestrial ecosystem. *Ecology* 81(5): 1402–1414.

Cebrian, J. 1999. Patterns in the fate of production in plant communities. *The American Naturalist* 154:449–468.

Chapin, F., III, A. J. Bloom, C. B. Field, and R. H. Waring. 1987. Plant responses to multiple environmental factors. *BioScience* 37 (1): 49–57.

Chapin, F. S., III, H. L. Reynolds, C. M. D'Antonio, and V. M. Eckhart. 1996. The functional role of species in terrestrial ecosystems. In B. Walker and W. Steffan, eds., *Global Change and Terrestrial Ecosystems*, 403–428. Cambridge: Cambridge University Press.

Chapin, F. S., III, O. E. Sala, I. C. Burke, J. P. Grime, D. U. Hooper, W. K. Lauenroth, A. Lombard, H. A. Mooney, A. R. Mosier, S. Naeem, S. W. Pacala, J. Roy, W. L. Steffen, and D. Tilman. 1998. Ecosystem consequences of changing biodiversity. *BioScience* 48:45–52.

Chesson, P. 1994. Multispecies competition in variable environments. *Theoretical Population Biology* 45:227–276.

Chesson, P. L. 1984. The storage effect in stochastic population models. *Lecture Notes in Biomathematics* 54:76–89.

———. 1985. Coexistence of competitors in spatially and temporally varying environments: a look at the combined effects of different sorts of variability. *Theoretical Population Biology* 28:263–287.

———. 1986. Environmental variation and the coexistence of species. In J. Diamond and T. Case, eds., *Community Ecology*, 240–256. New York: Harper and Row.

———. 1989. A general model of the role of environmental variability in communities of competing species. *Lectures on Mathematics in the Life Sciences* 20:97–123.

———. 1998. Spatial scales in the study of reef fishes: a theoretical perspective. *Australian Journal of Ecology* 23:209–215.

———. 2000a. General theory of competitive coexistence in spatially-varying environments. *Theoretical Population Biology* 58(3): 211–237.

———. 2000b. Mechanisms of maintenance of species diversity. *Annual Review of Ecology and Systematics* 31:343–366.

Chesson, P., and N. Huntly. 1997. The roles of harsh and fluctuating conditions in the dynamics of ecological communities. *The American Naturalist* 150 (5): 519–553.

Chesson, P. L., and R. R. Warner. 1981. Environmental variability promotes coexistence in lottery competitive systems. *The American Naturalist* 117 (6): 923–943.

Clein, J. S., and J. P. Schimel. 1995. Nitrogen turnover and availability during succession from alder to poplar in Alaskan taiga forests. *Soil Biology and Biochemistry* 27 (6): 743–752.

Cody, M. L., and H. A. Mooney. 1978. Convergence versus nonconvergence in Mediterranean-climate ecosystems. *Annual Review of Ecology and Systematics* 9:265–321.

Comins, H. N., and I. R. Noble. 1985. Dispersal, variability, and transient niches: species coexistence in a uniformly variable environment. *The American Naturalist* 126:706–723.

Costanza, R., R. d'Arge, R. deGroot, S. Farber, M. Grasso, B. Hannon, K. Limburg, S. Naeem, R. V. O'Neill, J. Paruelo, R. G. Raskin, P. Sutton, and M. Van den Belt. 1997. The value of the world's ecosystem services and natural capital. *Nature* 387:253–260.

Crawley, M. J. 1993. *GLIM for Ecologists*. London: Blackwell Scientific Publications.

Crawley, M. J., S. L. Brown, M. S. Heard, and G. R. Edwards. 1999. Invasion resistance in experimental grassland communities: species richness or species identity? *Ecology Letters* 2:140–148.

Crawley, M. J., and R. M. May. 1987. Population dynamics and plant community structure: competition between annuals and perennials. *Journal of Theoretical Biology* 125:475–489.

Crozier, R. H. 1997. Preserving the information content of species: genetic diversity, phylogeny, and conservation worth. *Annual Review of Ecology and Systematics* 28:243–268.

Cyr, H., and M. L. Pace. 1993. Magnitude and patterns of herbivory in aquatic and terrestrial ecosystems. *Nature* 361:148–150.

Daily, G. C., ed. 1997. *Nature's Services: Societal Dependence on Natural Ecosystems*. Washington, D.C.: Island Press.

Darwin, C. 1859. *On the Origin of Species by Means of Natural Selection*. New York: The Modern Library, Random House.

Dawes, E. A. 1989. Starvation, survival and energy reserves. In M. Fletcher and G. D. Floodgate, eds., *Bacteria in Their Natural Environments*, 43–79. Orlando: Academic Press.

De Angelis, D. L. 1975. Stability and connectance in food web models. *Ecology* 56:238–243.

———. 1992. *Dynamics of Nutrient Cycling and Food Webs*. London: Chapman and Hall.

DeBach, P. 1974. *Biological Control by Natural Enemies*. New York: Cambridge University Press.

REFERENCES

Del Giorgio, P. A., and J. M. Gasol. 1995. Biomass distribution in freshwater plankton communities. *American Naturalist* 146:135–152.

De Mazancourt, C., M. Loreau, and L. Abbadie. 1998. Grazing optimization and nutrient cycling: when do herbivores enhance plant production? *Ecology* 79:2242–2252.

Dendooven, L., E. Pemberton, and J. M. Anderson. 1996. Denitrification potential and reduction enzymes dynamics in a Norway spruce plantation. *Soil Biology and Biochemistry* 28 (2): 151–157.

De Ruiter, P. C., A. Neutel, and J. C. Moore. 1995. Energetics, patterns of interaction strengths, and stability in real ecosystems. *Science* 269:1257–1260.

Diaz, S., M. Cabido, and F. Casanoves. 1999. Functional implications of trait-environment linkages in plant communities. In E. Weiher and P. Keddy, eds., *Ecological Assembly Rules*, 338–362. Cambridge: Cambridge University Press.

Diehl, S. 1993. Relative consumer sizes and strengths of direct and indirect interaction in omnivorous feeding relationships. *Oikos* 68:151–157.

Diemer, M., J. Joshi, C. Körner, B. Schmid, and E. Spehn. 1997. An experimental protocol to assess the effects of plant diversity on ecosystem functioning utilized in a European research network. *Bulletin of the Geobotanical Institute ETH* 63:95–107.

Doak, D. F., D. Bigger, E. K. Harding, M. A. Marvier, R. E. Omalley, and D. Thomson. 1998. The statistical inevitability of stability-diversity relationships in community ecology. *The American Naturalist* 151:264–276.

Dodd, M. E., J. Silvertown, K. McConway, J. Potts, and M. Crawley. 1994. Stability in the plant communities of the Park Grass Experiment: the relationships between species richness, soil pH and biomass variability. *Philosophical Transactions of the Royal Society of London B Biological Sciences* 346:185–193.

Donald, C. M. 1951. Competition among pasture plants. I. Intraspecific competition among annual pasture plants. *Australian Journal of Agricultural Research* 2:355–376.

Duelli, P. 1997. Biodiversity evaluation in agricultural landscapes: an approach at two different scales. *Agriculture, Ecosystems and Environment* 62:81–91.

Dunbar, J., S. Takala, S. M. Barns, J. A. Davis, and C. R. Kuske. 1999. Levels of bacterial community diversity in four arid soils compared by cultivation and 16s rRNA gene cloning. *Applied and Environmental Microbiology* 65 (4): 1662–1669.

REFERENCES

Dutilleul, P. 1993. Spatial heterogeneity and the design of ecological field experiments. *Ecology* 74:1646–1658.

Dykhuizen, D. E. 1998. Santa Rosalia revisited: why are there so many species of bacteria? *Antonie van Leeuwenhoek* 73:25–33.

Edwards, P. J., and C. Abivardi. 1998. The value of biodiversity: where ecology and economy blend. *Biological Conservation* 83: 239–246.

Ehrlich, P. R., and A. H. Ehrlich. 1981. *Extinction: The Causes and Consequences of the Disappearance of Species.* New York: Random House.

———. 1992. The value of biodiversity. *Ambio* 21:219–226.

Elliot, E. T., L. G. Castañares, D. Perlmutter, and K. G. Porter. 1983. Trophic-level control of production and nutrient dynamics in an experimental planktonic community. *Oikos* 41:7–16.

Elton, C. S. 1958. *The Ecology of Invasions by Plants and Animals.* London: Methuen & Co. Ltd.

Estes, J. A., and D. O. Duggins. 1995. Sea otters and kelp forests in Alaska: generality and variation in a community ecological paradigm. *Ecological Monographs* 65:75–100.

Ewel, J. J., and S. W. Bigelow. 1996. Plant life-forms and tropical ecosystem functioning. In G. H. Orians, R. Dirzo, and J. H. Cushman, eds., *Biodiversity and Ecosystem Processes in Tropical Forests,* 101–126. New York: Springer-Verlag.

Ewel, J. J., M. J. Mazzarino, and C. W. Berish. 1991. Tropical soil fertility changes under monocultures and successional communities of different structure. *Ecological Applications* 1:289–302.

Farnsworth, K. D., and A. W. Illius. 1998. Optimal diet choice for large herbivores: an extended contingency model. *Functional Ecology* 12:74–81.

Fenchel, T., G. M. King, and T. H. Blackburn. 1998. *Bacterial Biogeochemistry: The Ecophysiology of Mineral Cycling.* San Diego: Academic Press.

Finlay, B. J., and G. F. Esteban. 1998. Freshwater protozoa: biodiversity and ecological function. *Biodiversity and Conservation* 7:1163–1186.

Finlay, B. J., S. C. Maberly, and J. I. Cooper. 1997. Microbial diversity and ecosystem function. *Oikos* 80:209–213.

Fischer, M., and D. Matthies. 1998. RAPD variation in relation to population size and plant fitness in the rare *Gentianella germanica* (Gentianaceae). *American Journal of Botany* 85:811–819.

Frank, D. A., and S. J. McNaughton. 1991. Stability increases with diversity in plant communities: empirical evidence from the 1988 Yellowstone drought. *Oikos* 62:360–362.

337

Gardner, M. R., and W. R. Ashby. 1970. Connectance of large dynamic (cybernetic) systems: critical values for stability. *Nature* 228:784.

Garnier, E., M.-L. Navas, M. P. Austin, J. M. Lilley, and R. M. Gifford. 1997. A problem for biodiversity-productivity studies: how to compare the productivity of multispecific plant mixtures to that of monocultures? *Acta Oecologica* 18:657–670.

Giller, K. E., M. H. Beare, P. Lavelle, A.-M. N. Izac, and M. J. Swift. 1997. Agricultural intensification, soil biodiversity and agroecosystem function. *Applied Soil Ecology* 6:3–16.

Ginnett, T. F. and M. W. Demment. 1995. The functional response of herbivores: analysis and test of a simple mechanistic model. *Functional Ecology* 9:376–384.

Gitay, H., J. B. Wilson, and W. G. Lee. 1996. Species redundancy: a redundant concept? *Journal of Ecology* 84:121–124.

Givnish, T. J. 1994. Does diversity beget stability? *Nature* 371:113–114.

Gleason, H. A. 1926. The individualistic concept of the plant association. *Bulletin of the Torrey Botanical Club* 53:7–26.

Goldstein, P. Z. 1999. Functional ecosystems and biodiversity buzzwords. *Conservation Biology* 13:247–255.

Goodman, D. 1975. The theory of diversity-stability relationships in ecology. *Quarterly Review of Biology* 50:237–267.

Grace, J. B. 1999. The factors controlling species density in herbaceous plant communities: an assessment. *Perspectives in Plant Ecology, Evolution and Systematics* 2:1–28.

Griffiths, B. S., K. Ritz, and R. E. Wheatley. 1997. Relationship between functional diversity and genetic diversity in complex microbial communities.In H. Insam and A. Rangger, eds., *Microbial Communities: Functional versus Structural Approaches*, 1–9. Berlin: Springer.

Grime, J. P. 1973. Competitive exclusion in herbaceous vegetation. *Nature* 242:344–347.

———. 1979. *Plant Strategies and Vegetation Processes.* Chichester: John Wiley and Sons.

———. 1997. Biodiversity and ecosystem function: the debate deepens. *Science* 277:1260–1261.

———. 1998. Benefits of plant diversity to ecosystems: immediate, filter and founder effects. *Journal of Ecology* 86:902–910.

Grover, J. P. 1997. *Resource Competition.* London: Chapman and Hall.

Grover, J. P., and R. D. Holt. 1997. Plants in trophic webs. In

REFERENCES

M. Crawley, ed., *Plant Ecology*, 556–567. London: Chapman and Hall.

——. 1998. Disentangling resource and apparent competition: Realistic models for plant-herbivore communities. *Journal of Theoretical Biology* 191:353–376.

Grover, J. P., and M. Loreau. 1996. Linking communities and ecosystems: trophic interactions as nutrient cycling pathways. In M. E. Hochberg, J. Clobert, and R. Barbault, eds., *Aspects of the Genesis and Maintenance of Biological Diversity*, 180–193. Oxford: Oxford University Press.

Groves, C., L. Valutis, D. Vosick, B. Neely, K. Wheaton, J. Touval, and B. Runnels. 2000. *Designing a Geography of Hope: A Practitioner's Handbook for Ecoregional Conservation Planning*. Arlington, Va.: The Nature Conservancy.

Grubb, P. 1977. The maintenance of species-richness in plant communities: the importance of the regeneration niche. *Biological Review* 52:107–145.

Gulledge, J., and J. P. Schimel. 1998. Moisture control over atmospheric CH_4 consumption and CO_2 production in diverse Alaskan soils. *Soil Biology and Biochemistry* 30 (8/9): 1127–1132.

Gulmon, S. L., and Mooney, H. A. 1986. Costs of defense on plant productivity. In T. J. Givnish, ed., *On the Economy of Plant Form and Function*, 681–698. Cambridge: Cambridge University Press.

Guterman, L. 2000. Have ecologists oversold biodiversity? *Chronicle of Higher Education*, October 13, 2000:A24–A26.

Hacker, S. D., and S. D. Gaines. 1997. Some implications of direct positive interactions for community species diversity. *Ecology* 78:1990–2003.

Hairston, N. G., Jr., and N. G. Hairston, Sr. 1993. Cause-effect relationships in energy flow, trophic structure, and interspecific interactions. *The American Naturalist* 142:379–411.

Hairston, N.G.S., F. E. Smith, and L. B. Slobodkin. 1960. Community structure, population control, and competition. *The American Naturalist* 106:249–257.

Halverson, L. J., T. M. Jones, and M. K. Firestone. 2000. Release of intracellular solutes by four soil bacteria exposed to dilution stress. *Soil Science Society of America Journal* 64 (5): 1630–1637.

Hamond, P. M. 1992. Species inventory. In B. Groombridge, ed., *Global Biodiversity, Status of the Earth's Living Resources*, 17–39. London: Chapman and Hall.

Hampicke, U. 1994. Ethics and economics of conservation. *Biological Conservation* 67:219–231.

Harder, W., and L. Dijkhuizen. 1983. Physiological responses to nutrient limitation. *Annual Review of Microbiology* 37:1–23.

Harper, J. L. 1977. *Population Biology of Plants.* London: Academic
Press.

Harrop, S. R. 1999. Conservation regulation: a backward step for
biodiversity? *Biodiversity and Conservation* 8:679–707.

Harte, J., and A. P. Kinzig. 1993. Mutualism and competition between plants and decomposers: implications for nutrient allocation in ecosystems. *The American Naturalist* 141:829–846.

Hastings, A. 1980. Disturbance, coexistence, history, and competition for space. *Theoretical Population Biology* 18:363–373.

Hector, A. 1998. The effect of diversity on productivity: detecting
the role of species complementarity. *Oikos* 82:597–599.

Hector, A., B. Schmid, C. Beierkuhnlein, M. C. Caldeira, M. Diemer, P. G. Dimitrakopoulos, J. Finn, H. Freitas, P. S. Giller,
J. Good, R. Harris, P. Högberg, K. Huss-Danell, J. Joshi, A. Jumpponen, C. Körner, P. W. Leadley, M. Loreau, A. Minns, C.P.H.
Mulder, G. O'Donovan, S. J. Otway, J. S. Pereira, A. Prinz, D. J.
Read, M. Scherer-Lorenzen, E.-D. Schulze, A.-S. D. Siamantziouras, E. M. Spehn, A. C. Terry, A. Y. Troumbis, F. I. Woodward,
S. Yachi, and J. H. Lawton, 1999. Plant diversity and productivity
experiments in European grasslands. *Science* 286:1123–1127.

Hector, A., A. J. Beale, S. J. Otway, and J. H. Lawton. 2000. Consequences of the reduction of plant diversity for litter decomposition: effects through litter quality and microenvironment. *Oikos*
90:357–371.

Hobbie, S. E. 1992. Effects of plant species on nutrient cycling.
Trends in Ecology and Evolution 7:336–339.

Hobbs, R. J., and L. F. Huenneke. 1992. Disturbance, diversity, and
invasion: implications for conservation. *Conservation Biology* 6:
324–337.

Hodgson, J. G., K. Thompson, P. J. Wilson, and A. Bogaard. 1998.
Does biodiversity determine ecosystem function? The Ecotron
experiment reconsidered. *Functional Ecology* 12:843–848.

Holden, P. A. 1995. The effects of water potential on the biodeg-
radation of a volatile hydrocarbon. Department of Soil Science,
University of California, Berkeley, Doctoral Dissertation.

Holdgate, M. 1996. The ecological significance of biological diversity. *Ambio* 25:409–416.

Holmes, P. M., and D. M. Richardson. 1999. Protocols for restoration based on recruitment dynamics, community structure, and
ecosystem function: perspectives from South African fynbos. *Restoration Ecology* 7:215–230.

REFERENCES

Holt, R. D. 1993. Ecology at the mesoscale: the influence of regional processes on local communities. In R. Ricklefs and D. Schluter, eds. *Species Diversity in Ecological Communities*, 77–88. Chicago: University of Chicago Press.

———. 1996. Food webs in space: An island biogeographic perspective. In G. A. Polis and K. O. Winemiller, eds., *Food Webs: Contemporary Perspectives*, 313–323. London: Chapman and Hall.

———. 1997. Community modules. In A. C. Gange and V. K. Brown, eds., *Multitrophic Interactions in Terrestrial Systems*, 333–350. Oxford: Blackwell Publishing.

———. [n.d.] Implications of system openness for local community structure and ecosystem function. In G. A. Polis, M. E. Power, and G. R. Huxel, eds., *Food Webs at the Landscape Level*. Chicago: University of Chicago, in press.

Holt, R. D., and A. Gonzalez. Manuscript. Implications of red noise for the structure of open ecological communities.

Holt, R. D., J. P. Grover, and D. Tilman. 1994. Simple rules for interspecific dominance in systems with exploitative and apparent competition. *The American Naturalist* 144:741–771.

Hooper, D. U. 1998. The role of complementarity and competition in ecosystem responses to variation in plant diversity. *Ecology* 79:704–719.

Hooper, D. U., and P. M. Vitousek. 1997. The effects of plant composition and diversity on ecosystem processes. *Science* 277:1302–1305.

———. 1998. Effects of plant composition and diversity on nutrient cycling. *Ecological Monographs* 68:121–149.

Horn, H. S., and R. H. MacArthur. 1972. Competition among fugitive species in a harlequin environment. *Ecology* 53:749–752.

Hubbell, S. P. 1979. Tree dispersion, abundance, and diversity in a tropical dry forest. *Science* 203:1299–1309.

———. 2001. *The Unified Neutral Theory of Biodiversity and Biogeography*. Princeton, N.J.: Princeton University Press.

Huisman, J., and F. J. Weissing. 1999. Biodiversity of plankton by species oscillations and chaos. *Nature* 402:407–410.

Hunt, H. W., E. R. Ingham, D. C. Coleman, E. T. Elliott, and C.P.P. Reid. 1988. Nitrogen limitation of production and decomposition in prairie, mountain meadow, and pine forest. *Ecology* 69 (4): 1009–1016.

Huntly, N. 1991. Herbivores and the dynamics of communities and ecosystems. *Annual Review of Ecology and Systematics* 22:477–503.

Hurd, L. E., and L. L. Wolf. 1974. Stability in relation to nutrient enrichment in arthropod consumers of old-field successional ecosystems. *Ecological Monographs* 44:465–482.

Huston, M. A. 1979. A general hypothesis of species diversity. *American Naturalist* 113:81–101.

———. 1994. *Biological Diversity: The Coexistence of Species on Changing Landscapes.* Cambridge: Cambridge University Press.

———. 1997. Hidden treatments in ecological experiments: reevaluating the ecosystem function of biodiversity. *Oecologia* 110:449–460.

Hutchinson, G. E. 1978. *An Introduction to Population Ecology.* New Haven: Yale University Press.

Huxel, G. R., and K. McCann. 1998. Food web stability: the influence of trophic flows across habitats. *The American Naturalist* 152:460–469.

Ives, A. R., K. Gross, and J. L. Klug. 1999. Stability and variability in competitive communities. *Science* 286:542–544.

Iwasa, Y., and J. Roughgarden. 1986. Interspecific competition among metapopulations with space-limited sub-populations. *Theoretical Population Biology* 30:194–214.

Jaksic, F. M., P. Feinsinger, and J. E. Jimenez. 1996. Ecological redundancy and long-term dynamics of vertebrate predators in semiarid Chile. *Conservation Biology* 10:252–262.

Jepson, W. L. 1993. *The Jepson Manual: Higher Plants of California.* Edited by J. C. Hickman. Berkeley: University of California Press. 1400pp.

Jolliffe, P. A. 1997. Are mixed populations of plant species more productive than pure stands? *Oikos* 80:595–602.

Jones, A. K. 1982. The interaction of algae and bacteria. In A. T. Bull and J. H. Slater, eds., *Microbial Interactions and Communities,* 189–247. London: Academic Press.

Jones, C. G., and J. H. Lawton, eds. 1995. *Linking Species and Ecosystems.* London: Chapman and Hall.

Jones, C. G., J. H. Lawton, and M. Shachak. 1997. Positive and negative effects of organisms as physical ecosystem engineers. *Ecology* 78:1946–1957.

Joshi, J., D. Matthies, and B. Schmid. 2000. Root hemiparasites and plant diversity in experimental grassland communities. *Journal of Ecology* 88:634–644.

Juhren, M., F. W. Went, and E. Phillips. 1956. Ecology of desert plants. IV. Combined field and laboratory work on germination in the Joshua Tree National Monument, California. *Ecology* 37:318–330.

Kaiser, J. 2000. Rift over biodiversity divides ecologists. *Science* 289:1282–1283.

REFERENCES

Killinbeck, K. T. 1996. Nutrients in senseced leaves: keys to the search for potential resorption and resorption proficiency. *Ecology* 77:1716–1727.

King, A. W. and S. L. Pimm. 1983. Complexity, diversity, and stability: a reconciliation of theoretical and empirical results. *The American Naturalist* 122:229–239.

Kinzig, A. P. 1994. *Mutualism and Competition Between Plants and Microorganisms: Implications for Nitrogen Allocation in Terrestrial Ecosystems.* Energy and Resources Group, University of California at Berkeley, Doctoral Dissertation.

Kinzig, A. P., S. Levin, S. Pacala, and J. Dushoff. 1999. Limiting similarity, species packing, and system stability for hierarchical competition-colonization models. *The American Naturalist* 153 (4): 371–383.

Kira, T., H. Ogawa, and K. Shinozake. 1953. Intraspecific competition among higher plants. 1. Competition-density-yield interrelationships in regularly dispersed populations. *Journal of the Institute of Polytechnics* (Osaka City University) D. 4:1–16.

Knight, B. P., S. P. McGrath, and A. M. Chaudri. 1997. Biomass carbon measurements and substrate utilization patterns of microbial populations from soils amended with cadmium, copper, or zinc. *Applied and Environmental Microbiology* 63 (1): 39–43.

Knops, J.M.H., D. Tilman, N. M. Haddad, S. Naeem, C. E. Mitchell, J. Haarstad, M. E. Ritchie, K. M. Howe, P. B. Reich, E. Siemann, and J. Groth. 1999. Effects of plant species richness on invasions dynamics, disease outbreaks, insect abundances and diversity. *Ecology Letters* 2 (5): 286–293.

Koenig, W. D., and J. Haydock. 1999. Oaks, acorns, and the geographical ecology of acorn woodpeckers. *Journal of Biogeography* 26:159–165.

Laakso, J., and H. Setälä. 1999. Sensitivity of primary production to changes in the architecture of belowground food webs. *Oikos* 87:57–64.

Lawler, S. P., and P. J. Morin. 1993. Food web architecture and population dynamics in laboratory microcosms of protists. *The American Naturalist* 141:675–686.

Lawton, J. H. 1994. What do species do in ecosystems? *Oikos* 71:367–374.

———. 1999. Biodiversity and ecosystem processes: theory, achievements and future directions. In M. Kato, ed., *The Biology of Biodiversity*, 119–131. Tokyo: Springer-Verlag.

Lawton, J. H., and V. K. Brown. 1993. Redundancy in ecosystems.

In E. D. Schulze and H. A. Mooney, eds., *Biodiversity and Ecosystem Functioning*, 255–270. New York: Springer-Verlag.

Lawton, J. H., S. Naeem, L. J. Thompson, A. Hector, and M. J. Crawley. 1998. Biodiversity and ecosystem function: getting the Ecotron experiment in its correct context. *Functional Ecology* 12:848–852.

Lehman, C. L., and D. Tilman. 2000. Biodiversity, stability, and productivity in competitive communities. *The American Naturalist*, 156 (5): 534–552.

Leibold, M. A. 1996. A graphical model of keystone predators in food webs: trophic regulation of abundance, incidence and diversity patterns in communities. *The American Naturalist* 147:784–812.

Leibold, M. A., and H. M. Wilbur. 1992. Interactions between food-web structure and nutrients on pond organisms. *Nature* 360:341–343.

Lennartsson, T., J. Tuomi, and P. Nilsson. 1997. Evidence for an evolutionary history of overcompensation in the grassland biennal *Gentianella campestris* (Gentianaceae). *The American Naturalist* 149:1147–1155.

Leopold, A. 1949. *A Sand County Almanac: Sketches Here and There.* New York: Oxford University Press.

Leps, J., J. Osbornova-Kosinova, and M. Rejmanek. 1982. Community stability, complexity and species life history strategies. *Vegetatio* 50:53–63.

Levin, S. A. 1974. Dispersion and population interactions. *The American Naturalist* 108:207–228.

———. 1992. The problem of pattern and scale in ecology. *Ecology* 73:1943–1967.

———. 1999. *Fragile Dominions: Complexity and the Commons.* Reading, Mass.: Perseus Books.

Levins, R., and D. Culver. 1971. Regional coexistence of species and competition between rare species. *Proceedings of the National Academy of Sciences of the United States of America* 68:1246–1248.

Lindeman, R. E. 1942. The trophic dynamic aspect of ecology. *Ecology* 23:399–418.

Loreau, M. 1992. Time scale of resource dynamics, and coexistence through time partitioning. *Theoretical Population Biology* 41:401–412.

———. 1994. Material cycling and the stability of ecosystems. *The American Naturalist* 143:508–513.

———. 1995. Consumers as maximizers of matter and energy flow in ecosystems. *American Naturalist* 145:22–42.

———. 1998a. Biodiversity and ecosystem functioning: a mechanistic model. *Proceedings of the National Academy of Sciences of the United States of America* 95:5632–5636.

————. 1998b. Separating sampling and other effects in biodiversity experiments. *Oikos* 82:600–602.

————. 1998c. Ecosystem development explained by competition within and between material cycles. *Proceedings of the Royal Society of London Series B* 265:33–38.

Loreau M., and R. D. Holt. Manuscript. On the implications of system openness for the relationship between biodiversity and ecosystem functioning.

Loreau, M., and N. Mouquet. 1999. Immigration and the maintenance of local species diversity. *The American Naturalist* 154:427–440.

Loria, M., and I. Noy-Meir. 1979–1980. Dynamics of some annual populations in a desert loess plain. *Israel Journal of Botany* 28:211–225.

MacArthur, R. H. 1955. Fluctuations of animal populations and a measure of community stability. *Ecology* 36:533–536.

MacArthur, R. H., and R. Levins, 1967. The limiting similarity, convergence and divergence of coexisting species. *The American Naturalist* 101:377–385.

MacArthur, R. H., and E. O. Wilson. 1967. *The Theory of Island Biogeography*. Princeton, N.J.: Princeton University Press.

MacNally, R., and G. P. Quinn. 1998. Symposium introduction: the importance of scale in ecology. *Australian Journal of Ecology* 23:1–7.

Margalef, R. 1969. Diversity and stability: a practical proposal and a model of interdependence. *Brookhaven Symposium on Biology* 22:25–37.

Marks, P. L. 1974. The role of pin cherry (*Prunus pennsylvanica* L.) in the maintenance of stability in northern hardwood ecosystems. *Ecological Monographs* 44:73–88.

May, R. M. 1972. Will a large complex system be stable? *Nature* 238:413–414.

————. 1974. *Stability and Complexity in Model Ecosystems*, 2nd ed. Princeton, N.J.: Princeton University Press. 265 pp.

————. 1975. Patterns of species abundance and diversity. In M. L. Cody and J. M. Diamond, eds., *Ecology and Evolution of Communities*, 81–120. Cambridge: Harvard University Press.

May, R. M., and M. A. Nowak. 1994. Superinfection, metapopulation dynamics, and the evolution of diversity. *Journal of Theoretical Biology* 170:95–114.

McGill, W. B., H. W. Hunt, R. G. Woodmansee, and J. O. Reuss. 1981. Phoenix, a model of the dynamics of carbon and nitrogen

REFERENCES

in grassland soils. In F. Clark and T. Rosswall, eds., *Terrestrial Nitrogen Cycles*, 49–115. Stockholm: Ecological Bulletins.

McGrady-Steed, J., P. M. Harris, and P. J. Morin. 1997. Biodiversity regulates ecosystem predictability. *Nature* 390:162–165.

McNaughton, S. J. 1977. Diversity and stability of ecological communities: a comment on the role of empiricism in ecology. *The American Naturalist* 111:515–525.

———. 1978. Stability and diversity of ecological communities. *Nature* 274:251–253.

———. 1985. Ecology of a grazing ecosystem: the Serengeti. *Ecological Monographs* 55:259–294.

———. 1993. Biodiversity and function of grazing ecosystems. In E.-D. Schulze and H. A. Mooney, eds., *Biodiversity and Ecosystem Function*, 361–383. Berlin: Springer-Verlag.

McNaughton, S. J., F. F. Banyikwa, and M. M. McNaughton. 1997. Promotion of the cycling of diet-enhancing nutrients by African grazers. *Science* 278:1798–1800.

McNaughton, S. J., M. Oesterheld, D. A. Frank, and K. J. Williams. 1989. Ecosystem-level patterns of primary productivity and herbivory in terrestrial habitats. *Nature* 341:142–144.

Mellinger, M. V., and S. J. McNaughton. 1975. Structure and function of successional vascular plant communities in central New York. *Ecological Monographs* 45:161–182.

Meyer, M. C., M. W. Paschke, T. McLendon, and D. Price. 1998. Decreases in soil microbial function and functional diversity in response to depleted Uranium. *Journal of Environmental Quality* 27:1306–1311.

Meyer, O. 1993. Functional groups of microorganisms. In E.-D. Schulze and H. A. Mooney, eds., *Biodiversity and Ecosystem Function*, 67–96. Berlin: Springer-Verlag.

Mikola, J., and H. Setälä. 1998. Relating species diversity to ecosystem functioning: mechanistic backgrounds and experimental approach with a decomposer food web. *Oikos* 83:180–194.

Milchunas, D. G. and W. K. Lauenroth. 1993. Quantitative effects of grazing on vegetation and soils over a global range of environments. *Ecological Monographs* 63:327–366.

Mills, L. S., M. F. Soule, and D. F. Doak. 1993. The keystone-species concept in ecology and conservation. *Bioscience* 43:219–224.

Mitchell, C. E. 2001. Global environmental change foliar fungal plant disease: testing the potential for interactive effects in a grassland ecosystem. Ph.D. thesis, University of Minnesota.

Moorcroft, P. R., G. C. Hurtt, and S. W. Pacala. 2001. Scaling rules for vegetation dynamics: a new terrestrial biosphere model for global change studies. *Ecological Monographs*. (Submitted).

Muko, S., and Y. Iwasa. 2000. Species coexistence by permanent spatial heterogeneity in a lottery model. *Theoretical Population Biology* 57:273–284.

Mulder, C.P.H., J. Koricheva, K. Huss-Danell, P. Högberg, and J. Joshi. 1999. Insects affect relationships between plant species richness and ecosystem processes. *Ecology Letters* 2:237–246.

Murdoch, W. W., and A. Stewart-Oaten. 1989. Aggregation by parasitoids and predators: effects on equilibrium and stability. *The American Naturalist* 134:288–310.

Myers, N. 1996. Environmental services of biodiversity. *Proceedings of the National Academy of Sciences of the United States of America.* 93:2764–2769.

Myers, N., R. A. Mittermeier, C. G. Mittermeier, G.A.B. Da Fonseca, and J. Kent. 2000. Biodiversity hotspots for conservation priorities. *Nature* 403:853–858.

Nadelhoffer, K. J., A. E. Giblin, G. R. Shaver, and J. A. Laundre. 1991. Effects of temperature and substrate quality on element mineralization in six Arctic soils. *Ecology* 72 (1): 242–253.

Naeem, S. 1998. Species redundancy and ecosystem reliability. *Conservation Biology* 12:39–45.

Naeem, S., K. Håkansson, L. J. Thompson, J. H. Lawton, and M. J. Crawley. 1996. Biodiversity and plant productivity in a model assemblage of plant species. *Oikos* 76:259–264.

Naeem, S., D. Hahn, and G. Schuurman. 2000. Producer-decomposer codependency modulates biodiversity effects. *Nature* 403:762–764.

Naeem, S., and S. Li. 1997. Biodiversity enhances ecosystem reliability. *Nature* 390:507–509.

———. 1998. Consumer species richness and autotrophic biomass. *Ecology* 79:2603–2615.

Naeem, S., L. J. Thompson, S. P. Lawler, J. H. Lawton, and R. M. Woodfin. 1994. Declining biodiversity can alter the performance of ecosystems. *Nature* 368:734–737.

———. 1995. Empirical evidence that declining species diversity may alter the performance of terrestrial ecosystems. *Philosophical Transactions of the Royal Society of London, B.* 347:249–262.

Nee, S., and R. M. May. 1992. Dynamics of metapopulations: habitat destruction and competitive coexistence. *Journal of Animal Ecology* 61:37–40.

Niklaus, P. A., E. Kandeler, P. W. Leadley, B. Schmid, D. Tscherko, and C. Körner. 2001a. A link between plant diversity, elevated CO_2, and soil nitrate. *Oecologia* 127:540–548.

Niklaus, P. A., P. W. Leadley, B. Schmid, and C. Körner. 2001b. A long-term field study on biodiversity x elevated CO_2 interactions in grasslands. *Ecological Monographs* 71:341–356.

Nisbet, R. M., S. Diehl, W. G. Wilson, S. D. Cooper, D. D. Donalson, and K. Kratz. 1997. Primary productivity gradients and short-term population dynamics in open systems. *Ecological Monographs* 67:535–553.

Norberg, J. 2000. Resource-niche complementarity and auto-trophic compensation determines ecosystem-level responses to increased cladoceran species richness. *Oecologia* 122:264–272.

Nüsslein, K., and J. M. Tiedje. 1998. Characterization of the dominant and rare members of a young Hawaiian soil bacterial community with small-subunit ribosomal DNA amplified from DNA fractionated on the basis of its guanine and cytosine content. *Applied and Environmental Microbiology* 64 (4): 1283–1289.

———. 1999. Soil bacterial community shift correlated with change from forest to pasture vegetation in a tropical soil. *Applied and Environmental Microbiology* 65 (8): 3622–3626.

Odum, E. P. 1953. *Fundamentals of Ecology*. Philadelphia, Penn.: Saunders.

Oksanen, L., S. D. Fretwell, J. Arruda, and P. Miemel. 1981. Exploitation ecosystems in gradients of primary productivity. *The American Naturalist* 118:240–261.

Oksanen, L., and T. Oksanen. 2000. The logic and realism of the hypothesis of exploitation ecosystems. *The American Naturalist* 155:703–723.

Oksanen, T., M. E. Power, and L. Oksanen. 1995. Ideal free habitat selection and consumer-resource dynamics. *The American Naturalist* 146:565–585.

Olff, H., and M. E. Ritchie. 1998. Effects of herbivores on grassland plant diversity. *Trends in Ecology and Evolution* 13:261–265.

Osenberg, C. W., O. Sarnelle, and D. Goldberg. 2000. Meta-analysis in ecology: concepts, statistics and applications. *Ecology* 80:1103–1104.

Pacala, S. W. and S. A. Levin. 1997. Biologically generated spatial pattern and the coexistence of competing species. In D. Tilman and P. Kareiva, eds., *Spatial Ecology: The Role of Space in Population Dynamics and Interspecific Interactions*, 204–232. Princeton, N.J.: Princeton University Press.

Pacala, S. W., and M. Rees. 1998. Models suggesting field experiments to test two hypotheses explaining successional diversity. *The American Naturalist* 152:729–737.

Pace, M. L., J. J. Cole, S. R. Carpenter, and J. F. Kitchell. 1999. Trophic cascades revealed in diverse ecosystems. *Trends in Ecology and Evolution* 14:483–488.

Paine, R. T. 1966. Food web complexity and species diversity. *The American Naturalist* 100:65–75.

Panikov, N. S. 1999. Understanding and prediction of soil microbial community dynamics under global change. *Applied Soil Ecology* 11:161–176.

Parton, W. J., D. S. Ojima, and D. S. Schimel. 1996. Models to evaluate soil organic matter storage and dynamics. In M. R. Carter and B. A. Stewart, eds., *Structure and Organic Matter Storage in Agricultural Soils*, 421–443. Boca Raton, Fla.: CRC Press.

Parton, W. J., D. S. Schimel, C. V. Cole, and D. S. Ojima. 1987. Analysis of factors controlling soil organic levels in Great Plains grasslands. *Soil Science Society of America Journal* 51:1173–1179.

Pastor, J., J. D. Aber, C. A. McClaugherty, and J. M. Melillo. 1984. Aboveground production and N and P cycling along a nitrogen mineralization gradient on Blackhawk Island, Wisconsin. *Ecology* 65:256–268.

Pastor, J., and Y. Cohen. 1997. Herbivores, the functional diversity of plant species, and the cycling of nutrients in ecosystems. *Theoretical Population Biology* 51:165–179.

Paustian, K., E. Levine, W. M. Post, and I. M. Ryzhova. 1997. The use of models to integrate information and understanding of soil C at the regional scale. *Geoderma* 79:227–260.

Petchey, O. L., P. T. McPhearson, T. M. Casey, and P. J. Morin. 1999. Environmental warming alters food-web structure and ecosystem function. *Nature* 402:69–72.

Pillar, V. D. 1999. On the identification of optimal plant functional types. *Journal of Vegetation Science* 10:631–640.

Pimm, S. L. 1979. Complexity and stability: another look at MacArthur's original hypothesis. *Oikos* 33:351–357.

———. 1982. *Food Webs*. London: Chapman and Hall.

———. 1984. The complexity and stability of ecosystems. *Nature* 307:321–326.

Pimm, S. L., and J. H. Lawton. 1977. Number of trophic levels in ecological communities. *Nature* 268:329–331.

Polis, G. A., W. B. Anderson, and R. D. Holt. 1997. Towards an integration of landscape ecology and food web ecology: the dynamics of spatially subsidized food webs. *Annual Review of Ecology and Systematics* 28:289–316.

Polis, G. A., and D. R. Strong. 1996. Food web complexity and community dynamics. *The American Naturalist* 147:813–846.

Polis G. A., and K. O. Winemiller, eds. 1996. *Food Webs: Contemporary Perspectives.* London: Chapman and Hall.

Polz, M. F., and C. M. Cavanaugh. 1998. Bias in template-to-product ratios in multitemplate PCR. *Applied and Environmental Microbiology* 64 (10): 3724–3730.

Power, M. E., D. Tilman, J. A. Estes, B. A. Menge, W. J. Bond, L. S. Mills, G. Daily, J. C. Castilla, J. Lubchenko, and R. T. Paine. 1996. Challenges in the quest for keystones. *BioScience* 46:609–620.

Purvis, A., P. M. Agapow, J. L. Gittleman, and G. M. Mace. 2000. Nonrandom extinction and the loss of evolutionary history. *Science* 288:328–330.

Raich, J. W., and W. H. Schlesinger. 1992. The global carbon dioxide flux in soil respiration and its relationship to vegetation and climate. *Tellus B* 44B (2): 81–99.

Rastetter, E. B., L. Gough, A. E. Hartley, D. A. Herbert, K. J. Nadelhoffer, and M. Williams. 1999. A revised assessment of species redundancy and ecosystem reliability. *Conservation Biology* 13:440–443.

Rathcke, B., and E. P. Lacey. 1985. Phenological patterns of terrestrial plants. *Annual Review of Ecology and Systematics* 16:179–214.

Ray, C., and A. Hastings. 1996. Density dependence: are we searching at the wrong spatial scale? *Journal of Animal Ecology* 65:556–566.

Reader, R. J., S. D. Wilson, J. W. Belcher, I. Wisheu, P. A. Keddy, D. Tilman, E. C. Morris, J. B. Grace, J. B. McGraw, H. Olff, R. Turkington, E. Klein, Y. Leung, B. Shipley, R. Vanhulst, M. E. Johansson, C. Nilsson, J. Gurevitch, K. Grigulis, and B. E. Beisner. 1994. Plant competition in relation to neighbor biomass—an intercontinental study with Poa Pratensis. *Ecology* 75:1753–1760.

Ricklefs, R. E., and D. Schluter. 1993. Species diversity: Regional and historical influences. In R. E. Ricklefs and D. Schluter, eds. *Species Diversity in Ecological Communities,* 350–362. Chicago: University of Chicago Press.

Ritchie, M. E., D. Tilman, and J.M.H. Knops. 1998. Herbivore effects on plant and nitrogen dynamics in oak savanna. *Ecology* 79:165–177.

Rolston, H., III. 1998. Challenges in environmental ethics. In M. E. Zimmerman, J. B. Callicott, G. Sessions, K. J. Warren, and J. Clark, eds., *Environmental Philosophy: From Animal Rights to Radical Ecology,* 124–144. Upper Saddle River, N.J.: Prentice Hall.

Rosenzweig, M. L., and Z. Abramsky. 1993. How are diversity and productivity related? In R. E. Ricklefs and D. Schluter, eds., *Spe-*

cies Diversity in Ecological Communities: Historical and Geographical Perspectives, 52–65. Chicago: University of Chicago Press.

Rosenzweig, M. L. 1995. Species Diversity in Space and Time. Cambridge: Cambridge University Press.

Roughgarden, J. 1976. Theory of Population Genetics and Evolutionary Ecology: An Introduction. New York: Macmillan.

Roughgarden, J., and J. Diamond. 1986. Overview: the role of species interactions in community ecology. In J. Diamond and T. J. Case, eds., Community Ecology, 333–343. New York: Harper & Row.

Sankaran, M., and S. J. McNaughton. 1999. Determinants of biodiversity regulate compositional stability of communities. Nature 401:691–693.

Scherer-Lorenzen, M. 1999. Effects of plant diversity on ecosystem processes in experimental grassland communities. Bayreuther Forum Ökologie 75:1–195.

Schimel, J. 1995. Ecosystem consequences of microbial diversity and community structure. In F.S.I. Chapin and C. Koerner, eds., Arctic and Alpine Biodiversity: Patterns, Causes, and Ecosystem Consequences, 239–254. New York: Springer Verlag.

Schimel, J. P., and J. S. Clein. 1996. Microbial response to freeze-thaw cycles in tundra and taiga soil. Soil Biology and Biochemistry 28 (8): 1061–1066.

Schimel, J. P., and J. Gulledge. 1998. Microbial community structure and global trace gases. Global Change Biology 4:745–758.

Schindler, D. W. 1998. Replication versus realism: the need for ecosystem-scale experiments. Ecosystems 1:323–334.

Schläpfer, F., and B. Schmid. 1999. Ecosystem effects of biodiversity—a classification of hypotheses and exploration of empirical results. Ecological Applications 9:893–912.

Schläpfer, F., B. Schmid, and I. Seidl. 1999. Expert estimates about effects of biodiversity on ecosystem processes and services. Oikos 84:346–352.

Schlesinger, W. H. 1997. Biogeochemistry. 2d ed. San Diego, Calif.: Academic Press.

Schmitz, O. J., A. P. Beckerman, and S. Litman. 1997. Functional responses of adaptive consumers and community stability with emphasis on the dynamics of plant-herbivore systems. Evolutionary Ecology 11:773–784.

Schulze, E. D., and H. A. Mooney, eds. 1993. Biodiversity and Ecosystem Function. New York: Springer-Verlag.

Schwartz, M., C. A. Brigham, J. D. Hoeksema, K. G. Lyons, M. H. Mills, and P. J. van Mantgem. 2000. Linking biodiversity to ecosystem function: implications for conservation ecology. *Oecologia* 122:297–305.

Selvin, S. 1995. *Practical Biostatistical Methods*. Pacific Grove, Calif.: Duxbury Press.

Shmida, A., and S. Ellner. 1984. Coexistence of plant species with similar niches. *Vegetatio* 58:29–55.

Siemann, E., D. Tilman, J. Haarstad, and M. Ritchie. 1998. Experimental tests of the dependence of arthropod diversity on plant diversity. *The American Naturalist* 152:738–750.

Simberloff, D. 1988. The contribution of population and community biology to conservation science. *Annual Review of Ecology and Systematics* 19:473–511.

———. 1998. Flagships, umbrellas, and keystones: is single-species management passé in the landscape era? *Biological Conservation* 83:247–257.

Slater, J. H., and D. Godwin. 1980. Microbial adaptation and selection. In D. C. Ellwood et al., eds. *Contemporary Microbial Ecology*, 137–160. New York: Academic Press.

Smedes, G. W., and L. E. Hurd. 1981. An empirical test of community stability: resistance of a fouling community to a biological patch-forming disturbance. *Ecology* 62:1561–1572.

Smith, A., and P. C. Allcock. 1985. The influence of species diversity on sward yield and quality. *Journal of Applied Ecology* 22:185–198.

Soule, M. 1996. Are ecosystem processes enough? *Wild Earth* 6:59–60.

Spehn, E. M., J. Joshi, B. Schmid, J. Alphei, and C. Körner. 2000a. Plant diversity effects on soil heterotrophic activity in experimental grassland systems. *Plant and Soil* 224:217–230.

Spehn, E. M., J. Joshi, B. Schmid, M. Diemer, and C. Körner. 2000b. Aboveground resource use increases with plant species richness in experimental grassland ecosystems. *Functional Ecology* 14:326–337.

Sprent, J. I., and P. Sprent. 1990. *Nitrogen Fixing Organisms: Pure and Applied Aspects*. London: Chapman and Hall.

Stahl, D. A. 1995. Application of phylogenetically based hybridization probes to microbial ecology. *Molecular Ecology* 4:535–542.

Staley, J. T. 1999. Bacterial biodiversity: a time for place. *ASM News* 65 (10): 681–687.

Stark, J. M., and M. K. Firestone. 1995. Mechanisms for soil moisture effects on activity of nitrifying bacteria. *Applied and Environmental Microbiology* 61 (1): 218–221.

———. 1996. Kinetic characteristics of ammonium-oxidizer com-

REFERENCES

munities in a California oak woodland-annual grassland. *Soil Biology and Biochemistry* 28 (10/11): 1307–1317.

Steinberg, P. D., J. A. Estes, and F. C. Winter. 1995. Evolutionary consequences of food chain length in kelp forest communities. *Proceedings of the National Academy of Science of the United States of America* 92:8145–8148.

Sterner, R. W., A. Bajpai, and T. Adams. 1997. The enigma of food chain length: absence of theoretical evidence for dynamic constraints. *Ecology* 78:2258–2262.

Stocker, R., C. Körner, B. Schmid, P. A. Niklaus, and P. W. Leadley. 1999. A field study of the effects of elevated CO_2 and plant species diversity on ecosystem-level gas exchange in a planted calcareous grassland. *Global Change Biology* 5:95–105.

Stork, N. 1997. Measuring global biodiversity and its decline. In M. L. Reaka-Kudla, D. E. Wilson, and E. O. Wilson, eds., *Biodiversity II*, 41–68. Washington, D.C.: Island Press.

Sugai, S. F., and J. P. Schimel. 1993. Decomposition and biomass incorporation of 14C-labeled glucose and phenolics in taiga forest floor: effect of substrate quality, successional state, and season. *Soil Biology and Biochemistry* 25 (10): 1379–1389.

Sullivan, G., and J. B. Zedler. 1999. Functional redundancy among tidal marsh halophytes: a test. *Oikos* 84:246–260.

Suzuki, M. T., and S. J. Giovannoni. 1996. Bias caused by template annealing in the amplification of mixtures of 16s rRNA genes by PCR. *Applied and Environmental Microbiology* 62 (2): 625–630.

Swift, M. J., and J. M. Anderson. 1993. Biodiversity and ecosystem function in agricultural systems. In E. D. Schulze and H. A. Mooney, eds., *Biodiversity and Ecosystem Function*, 15–41. Berlin: Springer Verlag.

Symstad, A. J., D. Tilman, J. Willson, and J.M.H. Knops. 1998. Species loss and ecosystem functioning: effects of species identity and community composition. *Oikos* 81:389–397.

Taylor, L. R., and I. P. Woiwod. 1980. Temporal stability as a density-dependent species characteristic. *Journal of Animal Ecology* 49:209–224.

Thompson, J. N. 1999. Specific hypotheses on the geographic mosaic of coevolution. *The American Naturalist* 153 Suppl.:S1–S14.

Tiedje, J. M., S. Amsung-Brempong, K. Nusslein, T. L. Marsh, and S. J. Flynn. 1999. Opening the black box of microbial diversity. *Applied Soil Ecology* 13:109–122.

Tilman, D. 1982. *Resource Competition and Community Structure.* Princeton, N.J.: Princeton University Press.

―――. 1988. *Plant Strategies and the Dynamics and Structure of Plant Communities*. Princeton, N.J.: Princeton University Press. 360 pp.

―――. 1990a. Constraints and tradeoffs—toward a predictive theory of competition and succession. *Oikos* 58:3–15.

―――. 1990b. Mechanisms of plant competition for nutrients: the elements of a predictive theory of competition. In J. B. Grace and D. Tilman, eds., *Perspectives on Plant Competition*, 117–141. San Diego, Calif.: Academic Press.

―――. 1994. Competition and biodiversity in spatially structured habitats. *Ecology* 75:2–16.

―――. 1996. Biodiversity: population versus ecosystem stability. *Ecology* 77:350–363.

―――. 1997a. Distinguishing between the effects of species diversity and species composition. *Oikos* 80:185.

―――. 1997b. Biodiversity and ecosystem functioning. In G. Daily, ed., *Nature's Services: Societal Dependence on Natural Ecosystems*, 93–112. Washington, D.C.: Island Press.

―――. 1999a. Diversity and production in European grasslands. *Science* 286:1099–1100.

―――. 1999b. Ecological consequences of changes in biodiversity: a search for general principles. *Ecology* 80:1455–1474.

Tilman, D., and J. A. Downing. 1994. Biodiversity and stability in grasslands. *Nature* 367:363–365.

Tilman, D., J. Knops, D. Wedin, P. Reich, M. Ritchie, and E. Sieman. 1997. The influence of functional diversity and composition on ecosystem processes. *Science* 277:1300–1302.

Tilman, D., C. L. Lehman, and K. T. Thomson. 1997. Plant diversity and ecosystem productivity: theoretical considerations. *Proceedings of the National Academy of Science of the United States of America* 94:1857–1861.

Tilman, D., C. Lehman, and C. E. Bristow. 1998. Diversity-stability relationships: statistical inevitability or ecological consequence? *The American Naturalist* 151:277–282.

Tilman, D., R. M. May, C. L. Lehman, and M. A. Nowak. 1994. Habitat destruction and the extinction debt. *Nature* (London) 371:65–66.

Tilman, D., and S. Pacala. 1993. The maintenance of species richness in plant communities. In R. E. Ricklefs and D. Schulter, eds., *Species Diversity in Ecological Communities*, 13–25. Chicago: University of Chicago Press.

Tilman, D., and P. B. Reich, J. Knops, D. Wedin, T. Mielke, and C. Lehman. 2001. Diversity and productivity in a long-term grassland experiment. *Science*, in press.

REFERENCES

Tilman, D., D. Wedin, and J. Knops. 1996. Productivity and sustainability influenced by biodiversity in grassland ecosystems. *Nature* 379:718–720.

Torsvik, V., F. L. Daae, R. A. Sandaa, and L. Ovreas. 1998. Novel techniques for analysing microbial diversity in natural and perturbed environments. *Journal of Biotechnology* 64:53–62.

Torsvik, V., K. Salte, R. Sorheim, and J. Goksoyr. 1990. Comparison of phenotypic diversity and DNA heterogeneity in a population of soil bacteria. *Applied and Environmental Microbiology* 56 (3): 776–781.

Tracy, C. R., and P. F. Brussard. 1994. Preserving biodiversity: species in landscapes. *Ecological Applications* 42:205–207.

Trenbath, B. R. 1974. Biomass productivity of mixtures. *Advances in Agronomy* 26:177–210.

Troumbis, A. Y., and D. Memtas. 2000. Observational evidence that diversity may increase productivity in Mediterranean shrublands. *Oecologia* 125:101–108.

Troumbis, A. Y., P. G. Dimitrakopoulos, A.-S. D. Siamantziouras, and D. Memtas. (2000). Hidden diversity and productivity patterns in mixed Mediterranean grasslands. *Oikos* 90 (3): 549–559.

Trumbore, S. E., O. A. Chadwick, and R. Amundson. 1996. Rapid exchange between soil carbon and atmospheric carbon dioxide driven by temperature change. *Science* 272 (19 April): 393–396.

Van der Heijden, M.G.A., J. N. Klironomos, M. Ursic, P. Moutogolis, R. Streitwolf-Engel, T. Boller, A. Wiemken, and I. R. Sanders. 1998. Mycorrhizal fungal diversity determines plant biodiversity, ecosystem variability and productivity. *Nature* 396:69–72.

Van der Heijden, M.G.A., J. N. Klironomos, M. Ursic, P. Moutoglis, R. Streitwolf-Engel, T. Boller, A. Wiemken, and I. R. Sanders. 1999. "Sampling effect," a problem in biodiversity manipulation? A reply to D. A. Wardle. *Oikos* 87:408–410.

Verchot, L., E. Franklin, and J. Gilliam. 1998. Effects of agricultural runoff dispersion on nitrate reduction in forest filter zones. *Soil Science Society of America Journal* 62:1719–1724.

Vitousek, P. M., J. D. Aber, R. H. Howarth, G. E. Likens, P. A. Matson, D. W. Schindler, W. H. Schlesinger, and D. G. Tilman. 1997. Human alteration of the global nitrogen cycle: source and consequences. *Ecological Applications* 3:737–750.

Vitousek, P., C. M. Dantonio, L. L. Loope, and R. Westbrooks. 1996. Biological invasions as global environmental change. *American Scientist* 84:468–478.

Vitousek, P. M., and D. U. Hooper. 1993. Biological diversity and terrestrial ecosystem biogeochemistry. In E.-D. Schulze, and H. A. Mooney, eds., *Biodiversity and Ecosystem Function*, 3–14. Berlin: Springer-Verlag.

Vitousek, P. M., L. R. Walker, L. D. Whiteaker, D. Mueller-Dombois, and P. A. Matson. 1987. Biological invasion by *Myrica faya* alters ecosystem development in Hawaii. *Science* 238:802–804.

Von Foerster, H., 1959. Some remarks on changing populations. In F. Stohlman, Jr., ed., *The Kinetics of Cellular Proliferations*, 382–407. New York: Grune & Stratton.

Waide, R. B., M. R. Willig, C. F. Steiner, G. Mittelbach, L. Gough, S. I. Dodson, G. P. Juday, and R. Parmenter. 1999. The relationship between productivity and species richness. *Annual Review of Ecology and Systematics* 30:257–300.

Walker, B. 1995. Conserving biological diversity through ecosystem resilience. *Conservation Biology* 9:747–752.

Walker, B., A. Kinzig, and J. Langridge. 1999. Plant attribute diversity, resilience, and ecosystem function: the nature and significance of dominant and minor species. *Ecosystems* 2:95–113.

Wall, D. H., and J. C. Moore. 1999. Interactions underground. *BioScience* 49:109–117.

Wardle, D. A. 1999. Is "sampling effect" a problem for experiments investigating biodiversity-ecosystem function relationships? *Oikos* 87:403–407.

Wardle, D. A., K. I. Bonner, and K. S. Nicholson. 1997. Biodiversity and plant litter: experimental evidence which does not support the view that enhanced species richness improves ecosystem function. *Oikos* 79:247–258.

Wardle, D. A., and A. Ghani. 1995. Why is the strength of relationships between pairs of methods for estimating soil microbial biomass often so variable? *Soil Biology and Biochemistry* 27 (6): 821–828.

Wardle, D. A., M. A. Huston, J. P. Grime, F. Berendse, E. Garnier, W. K. Lauenroth, H. Setala, and S. D. Wilson. 2000. Biodiversity and ecosystem function: an issue in ecology. *The Ecological Society of America Bulletin* 81 (3): 232–235.

Wardle, D. A., and K. S. Nicholson. 1996. Synergistic effects of grassland plant species on soil microbial biomass and activity: implications for ecosystem-level effects of enriched plant diversity. *Functional Ecology* 10:410–416.

Wardle, D. A., O. Zackrisson, G. Hörnberg, and C. Gallet. 1997. The influence of island area on ecosystem properties. *Science* 277:1296–1299.

REFERENCES

Wedin, D., and D. Tilman. 1990. Species effects on nitrogen cycling: a test with perennial grasses. *Oecologia* 84:433–441.

———. 1993. Competition among grasses along a nitrate gradient: initial conditions and mechanisms of competition. *Ecological Monographs* 63 (2):199–229.

Weiher, E., and P. A. Keddy. 1995. Assembly rules, null models and trait dispersion: new questions from old patterns. *Oikos* 74:159–164.

Weiher, E., A. van der Werf, K. Thompson, M. Roderick, E. Garnier, and O. Eriksson. 1999. Challenging Theophrastus: a common core list of plant traits for functional ecology. *Journal of Vegetation Science* 10:609–620.

Whittaker, R. H. 1957. The kingdoms of the living world. *Ecology* 38:536–538.

———. 1959. On the broad classification of organisms. *Quarterly Review of Biology* 34:210–226.

———. 1975. *Communities and Ecosystems.* 2d ed. New York: Macmillan.

Whittaker, R. H., and L. Margulis. 1978. Protist classification and the kingdoms of organisms. *BioSystems* 10:3–18.

Wilson, E. O. 1992. *The Diversity of Life.* Cambridge, Mass.: Harvard University Press, Belknap Press.

Wünsche, L., L. Brüggemann, and W. Babel. 1995. Determination of substrate utilization patterns of soil microbial communities: an approach to assess population changes after hydrocarbon pollution. *FEMS Microbiology Ecology* 17:295–306.

Yachi, S., and M. Loreau. 1999. Biodiversity and ecosystem productivity in a fluctuating environment: the insurance hypothesis. *Proceedings of the National Academy of Sciences of the United States of America* 96:57–64.

Zak, D. R., D. Tilman, R. R. Parmenter, C. W. Rice, F. M. Fisher, J. Vose, D. Milchunas, and C. W. Martin. 1994. Plant-production and soil-microorganisms in late-successional ecosystems—a continental-scale study. *Ecology* 75:2333–2347.

Zelles, L., R. Rackwitz, Q. Y. Bai, T. Beck, and F. Beese. 1995. Discrimination of microbial diversity by fatty acid profiles of phospholipids and lipopolysaccharides in differently cultivated soils. *Plant and Soil* 170:115–122.

Zheng, D. W., J. Bengtsson, and G. I. Ågren. 1997. Soil food webs and ecosystem processes: decomposition in donor-control and Lotka-Volterra systems. *The American Naturalist* 149:125–148.

.

Index

acorn woodpeckers, 44–45
adaptation, microbial, 270, 272–82
alien species, 311
allocation, 193, 205
animal rights movement, 296–97
annual plants, vs. perennials, 91
aquatic ecosystems, 103–10, 117, 140
aquatic microorganisms, 48, 103
arthropods, 118, 299
autotrophs: as class of trophic structure, 99; heterotrophs, interactions with (*see* heterotrophs); multilevel-trophic models and, 114–17

basal groups, 100
biocentric conservationism, 296–97
BIODEPTH project, 71–73, 89, 94; differences between locations, 74–75; experimental design, 145; inverse diversity–functioning relationship, 152, 162; nitrogen fixers, effect of, 81–82; productivity, multiple influences on, 74; related studies, comparisons with, 89–93; relationships within and between sites, 93–94; sampling effect models and productivity, 84–89; sampling effects, 82–84, 91–93, 133; species richness vs. composition, 79–81, 304; species richness vs. functional groups, 75–79
biodiversity, and ecosystem functioning . *See* diversity–functioning relationship
Biodiversity and Ecosystem Function, xiii
Biodiversity I, 53–64, 68, 90. *See also* Cedar Creek studies

Biodiversity II, 63, 68, 90, 92. *See also* Cedar Creek studies
biogeochemical models, 96–97, 169–74
biomass, 52; BIODEPTH project and (*see* BIODEPTH project); heterotrophics and autotrophic, 105–9; in a Mediterranean ecosystem, 240; successional niche hypothesis and, 179–83, 191–93; temporal stability and, 31–32; total community, 48–49, 52, 54–60; during the transition from sampling effects to niche complementarity, 154–66; trophic structure and, 103–4
biota: classes of functional groups, 98–100; classification and diversity of, 96–98. *See also* trophic structure
buffered population growth, 217, 219–20, 222–23

carbon consumption, 98–99
carbon cycling, 96, 98, 100–103, 112–13, 115–16
carbon storage: biodiversity and, 314–15; in plant biomass, 169–70, 178, 186, 188–93, 195–98, 203–4, 206, 208, 210–11, 213, 225, 235–36, 253–56, 288–89; in soil, 128, 171–74, 181–83
carbon uptake. *See* production
Cedar Creek studies: Biodiversity I, 53–64, 68, 90; Biodiversity II, 63, 68, 90, 92; biomass growth, 157; controversy over, xiii–xiv; deer exclosures, 246–47; diversity-stability hypothesis and, 45–48; undisturbed grasslands, 65–66
CENTURY model, 171–73

INDEX

Creek studies (*see* Cedar Creek studies); controversy over, xiv; designing studies, 14–15, 141, 144–48; diversity-productivity hypothesis, 49–52, 69 (*see also* production); diversity-stability hypothesis, 43–49, 69 (*see also* stability); Ecotron foodweb, xiv, 52; future studies, suggestions for, 148–50; sampling mechanisms vs. niche complementarity, 3–5, 68; trophic structure in research, 117–19, 140–43 diversity-nutrient retention hypothesis, 50

ecosystem engineers, 141
ecosystems: aquatic (*see* aquatic ecosystems); common model of (*see* common ecosystem model); forests (*see* forests); functioning of and biodiversity (*see* diversity–functioning relationship); grasslands (*see* grasslands); microbial ecology and, 269–70 (*see also* microbial communities); reliability, 108; variation in, and species specialization (*see* environmental niches)
Ecotron foodweb experiment, xiv, 52
endangered species, 296
environmental niches: aspects of, 215–16; ecosystem variation and species specialization, 213–15, 237–44; spatial, 222–26, 239, 241; spatio-temporal, 177, 222–23, 241; temporal, 216–22, 226–28, 239–41; temporal in a Mediterranean ecosystem, 228–37
Estes, J. A., 258
European grasslands, 65, 67. *See also* BIODEPTH project
evapotranspiration, 170, 186, 189, 191, 195–97, 202, 205, 207, 209–11, 225, 235–36, 253–56, 261

evolution, 213–15, 258–59, 272–74
extinction: calculation of probability of, 29–30; experimental prediction of, 144; human perturbations and, 307; local, 21, 107, 221, 257–58, 321; rainfall perturbation and, 234; random vs. nonrandom processes of, 149–50

food webs: dynamics of, 247, 249–50; Ecotron foodweb experiment, xiv, 52
forests: diversity of, 12–13; human disturbance of, 309; microbial communities and, 273–76, 281, 286; succession in, 175–76, 184–86
functional diversity, 12, 14, 145. *See also* diversity–functioning relationship
functional groups: BIODEPTH project, 73; classes of, 98–100; compositional effects and, 79–81; diversity and, 13–14, 48–52, 60–63, 260; dominant species and, 301; ecological differences among species and, 144–46; ecosystem functioning and, 69; implications of loss of, 310; in microcosm experiment, 107; pollution and, 309; productivity and, 89; species richness vs., 75–79. *See also* trophic structure

GCTE. *See* Global Change in Terrestrial Ecosystems program
global change, 213–15, 244, 269–70, 281–83, 285–86, 316
Global Change in Terrestrial Ecosystems (GCTE) program, xvi
grasslands: BIODEPTH project (*see* BIODEPTH project); Cedar Creek studies (*see* Cedar Creek studies); composition vs. diversity, 15–16, 52; diversity–productivity